Menu Collection

Problems Adapted from
Mathematics Teaching in the Middle School

Compiled and edited by

C. Patrick Collier

University of Wisconsin—Oshkosh
Oshkosh, Wisconsin

NATIONAL MIDDLE SCHOOL ASSOCIATION
Columbus, Ohio
NATIONAL COUNCIL OF TEACHERS OF MATHEMATICS
Reston, Virginia

Library of Congress Cataloging-in-Publication Data:

Collier, C. Patrick (Charles Patrick), 1937-
    Menu collection : problems adapted from Mathematics teaching in the middle school /
compiled and edited by C. Patrick Collier.
        p. cm,
    Includes index.
    ISBN 0-87353-490-5
        1. Mathematics—Problems, exercises, etc. 2. Mathematics—Study and teaching (Middle
school) I. Mathematics teaching in the middle school. II. Title.

QA43 .C23 2000
510'.76—dc21

00-045255

Printed in the United States of America

Contents

List of Menus

		Appetizers	Main Courses	Desserts	Total
Buffets and Smorgasbords					
1	Appetizer Buffet	8	0	0	8
2	Dessert Buffet I	0	0	6	6
3	Dessert Buffet II	0	0	6	6
4	Food for Thought	0	0	8	8
5	Main-Course Buffet I	0	8	0	8
6	Main-Course Buffet II	0	8	0	8
7	Smorgesbord I	2	3	2	8
8	Smorgasbord II	2	3	3	8
Theme Menus					
9	Calendar	2	4	1	7
10	Clocks	1	3	5	9
11	Counting	2	3	3	8
12	Cubes	2	1	5	8
13	Digits	3	3	2	8
14	Factors and Multiples I	3	4	1	8
15	Factors and Multiples II	2	4	2	8
16	Find a Number	3	3	2	8
17	Geometry	3	3	3	9
18	Max–Min	3	3	2	8
19	Measurement I	3	2	3	8
20	Measurement II	2	3	3	8
21	Patterns	3	3	2	8
22	Probability I	3	2	1	6
23	Probability II	3	3	1	7
24	Proportion	3	2	2	7
25	Races	1	3	2	6
26	Square Figures	0	4	4	8
27	Square Numbers	2	2	4	8
28	Sums I	2	3	2	7
29	Sums II	3	3	1	7
30	Triangles	1	5	2	8
TOTALS		62	88	78	228

Preface

As I was finishing a three-year term on the *MTMS* Editorial Panel, the Educational Materials Committee (EMC) was exploring the idea of publishing a collection of problems from *Mathematics Teaching in the Middle School* (*MTMS*). The *MTMS* Panel suggested that the collection retain the "menu" format so that it would be familiar to readers of the *MTMS*. Encouraged by my colleagues on the panel, I accepted the assignment to edit this publication.

I began by compiling a short history of the "Menu of Problems" and the "Food for Thought" departments of the *MTMS*. That history appears in this publication. Then I selected about 225 problems from the journal, gave each a name, and typed them into my computer. There followed the task of grouping the problems by topic or by theme. Eventually a number of themes emerged and menus were constructed about those themes. A number of problems remained that could not be grouped to fill out theme menus, so smorgasbords were created, along with a few menus that contained only appetizers, only main courses, or only desserts. The result was the collection of thirty menus that appear here.

Almost all the problems in this collection appear exactly as they were published in the *MTMS* "Menu of Problems" section. Each problem in this book contains a reference to the month, year, and problem number of its publication. (For example, the reference "[Feb. 1995 (4)]" indicates that the problem appeared in the *MTMS* menu as problem 4 in the February 1995 issue.)

The problems taken from the "Food for Thought" department were given names when they appeared in the *MTMS* (their citations in this collection also contain the month, year, and page number on which they appeared in the journal). However, none of the problems taken from the regular menu appeared with names; the names that appear here have been specifically added for this publication.

The only editing of problem statements involved selecting parts of problems that had multiple parts or combining two related problems into a single entry. Most of the solutions appear exactly as they were originally published. Nevertheless, some minor errors have been corrected, and a few solutions have been broadened to include another approach.

I had suggested that the publication include a brief introduction to problem solving and some ideas about how teachers had used the problems in their classrooms. The editorial panel of the EMC agreed, so a brief article on problem solving was written, reviewed, and rewritten. As the work progressed, it appeared to me that something on problem posing should be added, partly to highlight that important aspect of problem solving and partly to provide readers with a tool for extending the usefulness of this publication.

When this publication was originally planned, it was hoped that there would be an extensive section devoted to suggestions from teachers on how they had used the "Menu of Problems" in the *MTMS* and how one might use this problem collection in the middle-grades classroom. The solicitation of such suggestions produced only a few responses. Most of these responses reported how the "Menu of Problems" was used in "extra credit" situations or in a problem-solving competition. Although these are appropriate uses of the "Menu of Problems," it is hoped that the publication of this menu collection will encourage teachers to incorporate problem solving into their mainstream curricula. We wanted to encourage teachers to think of problem solving and problem posing as the means of teaching meaningful mathematics to everyone rather than as simply a new topic to be added to the curriculum or as a way of enriching the curriculum for only the better students. We have thus decided to include no explicit tips from teachers on how to use this publication.

Finally came the task of indexing and providing the resources for readers to locate problems by classification (appetizer, main course, or dessert), by topic, or by name. To make the collection more useful, I have included some specific suggestions on problem posing. The idea is that students should not consider themselves done with a problem when they find an answer, because that means they are done learning from the problem. To extend the opportunity to learn, students can formulate similar problems; they can generalize; they can create different contexts; they can be creative.

I hope this publication can be useful to middle school teachers in their quest to teach students the real and lifelong skills of problem solving. I hope, also, that it serves to increase attention for the *Mathematics Teaching in the Middle School* journal and for the "Menu of Problems" department in that journal.

Introduction

When the first editorial panel of *Mathematics Teaching in the Middle School* (*MTMS*) planned the structure of the journal, they decided to have features called *departments* that would appear in each issue or on a fairly regular basis. These departments are prepared by appointed editors and include articles that are submitted by classroom teachers and other authors.

One of the departments is a collection of problems called the "Menu of Problems." Each menu contains appetizers, main courses, and desserts. The appetizers are usually warm-up problems that are not very time-consuming and are suitable for the lower grade range of the grades covered by the journal. The main-course problems are usually a bit more time-consuming and more challenging. The desserts are even more challenging, more time-consuming, and more suitable for students who have had more experience in solving problems. Issues of MTMS that cover a single month, say May, have one menu of problems called the "May Menu of Problems." Issues of MTMS that cover two months, say October and November, have two menus, an "October Menu of Problems" and a "November Menu of Problems." In a typical year, there are nine menus.

The December 1998 issue of *MTMS* included the publication of the forty-second "Menu of Problems." Linda Walker was the initial editorial panel liaison and worked with David Stout from the University of West Florida for the first fifteen menus, which appeared from April of 1994 to December of 1995. Those menus contained a total of 273 problems: 96 appetizers, 120 main courses, and 57 desserts. There were about 18 problems in each menu, with about 6 appetizers, 8 main courses, and 4 desserts. One of the more noticeable changes that occurred during their tenure was the increased detail in the solutions. In the first several issues, solutions were brief and recorded in small print. In the later issues, the solutions were more detailed and recorded in print that was the same size as the rest of the journal.

C. Patrick Collier, from the University of Wisconsin—Oshkosh, took over as editorial panel liaison with the January 1996 menu and continued in that position for the next twenty-seven menus, concluding with the December 1998 menu. For the first nine menus (January 1996 to December 1996), the editors were Gerry Greer and Larry Wantuck of Broward County Public Schools, Florida. They produced a total of 162 problems, including 46 appetizers, 77 main courses, and 39 desserts. There were still about 18 problems in each menu, but the distribution changed—slightly fewer appetizers and more main courses and desserts. During this time, the solutions became even more detailed and frequently included alternative approaches. Also the menu changed physically in response to readers' requests that menus appear in the same orientation on the page as the rest of the journal.

For the last eighteen of the forty-two menus published between April 1994 and December 1998, the editors were Connie Laughlin from Steffen Middle

School in Mequon, Wisconsin, and Joseph Georgeson from Glen Hill Middle School in Glendale, Wisconsin. They produced 270 problems and continued the trend toward having as many desserts as appetizers. The average number of problems in a menu reached 15, with about 6 main courses and the rest divided equally between appetizers and desserts. They continued to produce solutions and alternative solutions that took up at least as much journal space as the problems.

From April 1994 to December 1998, the "Menu of Problems" department has posed just over 700 problems: about 300 have been main courses, about 220 have been appetizers, and about 180 have been desserts. During that entire span, attention has been given to publishing only problems that have been used in the middle grades by those who are the editors of the menu or problems that have been submitted by other middle-grades teachers.

The "Menu of Problems" department continues to be an important part of *MTMS* to this day. The current editor is Gretchen Muller of Del Mar Middle School in Tiburon, California.

Some Thoughts on Problem Solving

*T*his collection of problems from the "Menu of Problems" from *Mathematics Teaching in the Middle School* (*MTMS*) provides a resource for teachers to use in the classroom. We hope that middle school teachers will use this publication to improve the problem-solving experiences of their students. This article attempts to provide middle school teachers with some perspectives on problem solving. These perspectives have been collected over six years of experience with workshops on problem solving for middle school teachers.

We begin by describing a framework for organizing the problem-solving experience. There are advantages to thinking of the beginning, the middle, and the end phases of the problem-solving experience. We describe each of these phases and list some questions that help to define the phases as well as guide our thinking. There follows an example of a particular problem and a thought process that includes all three stages. Next, we provide a list of suggestions to improve problem solving. Finally, we discuss some of the issues associated with teaching problem solving.

ORGANIZING THINKING ABOUT PROBLEM SOLVING

The general approach to problem solving that is described in *Thinking Mathematically* by John Mason, with Leone Burton and Kaye Stacey (Reading, Mass.: Addison Wesley Publishing Co., 1982), is very useful for middle-grades students. This book describes this process as comprising three phases: *entry, attack,* and *review.* Acceptance of these three phases immediately establishes problem solving as a time-consuming activity. This counteracts the feelings of many students that you either know the answer to a given problem or you don't, and if you don't know, there isn't anything you can do. It is crucial to establish that if you have a problem, by definition of "problem," you do not initially know the answer or an immediate path to an answer. Problem solving is not an algorithmic process. There is no one set of directions that is guaranteed to yield a solution to all problems. There are some widely applicable strategies, but even if you know many of them, you are still destined to be *stuck* much of the time if you have a complicated problem to solve.

The lines between "entry," "attack," and "review" are not strictly drawn. Some problem-solving activities are clearly identified with one phase, to the exclusion of other phases. But some activities do not clearly fall in to one category or another. The purpose of identifying phases is not so much to analyze problem-solving stories as it is to provide a general framework for teaching and learning about problem solving.

The "entry" phase is the phase in which one attempts to understand the problem. Students frequently believe that this phase consists of reading the problem, perhaps more than once, and recalling the definitions of the words that appear in the problem statement. They tend to regard this phase as relatively unimportant. Experience teaches that the entry phase is most important. The earlier portion of the entry phase is sometimes called the *play* phase to reflect the idea that the activity is "playing" in the sense that you are not consciously trying to solve the problem or even trying to develop a plan. Rather, you are trying to gain experience with the problem setting, and you gain that experience by playing with the elements of the problem situation. The later portion of the entry phase contains activities that are more purposefully directed at finding a solution. However, that purposeful activity cannot occur until the problem situation is understood sufficiently to direct or motivate the activity.

The "attack" phase is the phase in which the solution is obtained or it is determined that one must return to the entry phase to get a better understanding of the problem situation. It is in this stage that one works more systematically, more deliberately, and more purposefully. Students typically have more familiarity with this phase because they usually spend the most time in it. It is in this phase that becoming "stuck" is more frustrating because there is an expectation that if the problem situation is sufficiently understood, the attack should be obvious. This may happen to students who oversimplify problem solving to "devise a plan" and "carry out the plan." In the solution of any substantial problem, a plan does not often emerge completely developed and ready to be carried out. Rather, a plan emerges slowly and incompletely, if at all. Frequently, you are not aware of a plan until you look back over a completed process and examine a sequence of activities.

The "review" phase is the phase where reflection and, consequently, most real learning takes place. Students typically view this stage as the time when they check with an authority to determine if their solutions are correct. This should be the stage where an argument is formulated that "proves" the solution is correct. (Students who ask, "Am I done yet?" need to learn that if they ask that question, they are not done. Being done presumes that you know you are done.) If an argument cannot be constructed, then it may be because the solution is incorrect, and the attempt to construct the argument may actually reveal what is incorrect about the solution. This is also the stage where students may construct a more "elegant" solution than they had at first and learn to appreciate multiple solutions and to identify solutions that are more elegant than others. It is the stage where we review what we learned about mathematics, about problem solving, and about ourselves as a result of the experience. It is an opportunity to understand that we can learn a great deal about mathematics, about problem solving, and about ourselves from the experience of working on a problem even when we fail to reach a correct solution to that problem.

Questions That Will Help Guide the Phases of Problem Solving

Some questions to ask during the ENTRY phase follow:
- What was given in the problem? What do I know?
- Do I need any other information? What more would I like to know? How would knowing that help?

- How do I interpret the problem? Is more than one interpretation possible? Is more than one likely? Can I choose one interpretation and continue?
- Can I organize what I know in a useful form—a table, a structured list, a graph, a picture, or a diagram?
- What am I asked to find? Can I make a wild guess at an answer? Can I make an educated estimate? If there are many answers, can I find one of the simple answers?
- How would I check an answer if I had one?
- Can I simplify the problem by working on a part of it or by ignoring one or more of the conditions? Can I find something to do?

Some questions to ask during the ATTACK phase include:

- Can I make a conjecture (a guess)? Can I check to see if the conjecture is true? If the conjecture is true, what does that tell me about the problem? If the conjecture is false, can I modify it so that it might be true?
- Can I construct some examples and look for a pattern?
- If I am "stuck," have I reviewed what I know? How I can use what I know? Have I tried to organize what I know in a more useful way? Have I reviewed what I want and how I might get what I want? Have I explicitly considered how to use what I know to get what I want?
- Have I considered these potential strategies: work backward, make a model, use algebraic representation, solve an easier related problem, employ the process of elimination, or guess and check and revise?
- Have I tried to change my point of view by trying to look at the problem in a different way?
- What have I tried that has not worked? Why did it not work? Would trying it again produce a different result?

Some questions to ask during the REVIEW phase follow:

- How do I know I have a solution? How could I convince a friend that my solution is correct?
- Can I find the solution in a different way? Is one way simpler and easier to present than another?
- What did I learn about mathematics while solving this problem?
- What did I learn about problem solving while solving this problem?
- What did I learn about myself while solving this problem?
- Can I make up another problem related to this problem that would be a good problem to solve?

An Illustration of the General Approach to Problem Solving

Consider the following problem:

Fifteen cookies are to be distributed to four children so that each of them has a different number of cookies. How many ways can this be done?

ENTRY thoughts. On the surface, the problem is easy to understand. I can represent the fifteen cookies by ○○○○○ ○○○○○ ○○○○○ and name the four children Alvin, Bret, Chas, and Delia. Let's see if I can find one way to distribute the cookies. Suppose we give 3 to Alvin, leaving 12 to distribute.

Let's give 4 to Bret, leaving 8 to distribute. I can represent what has been done so far by [OOO] [OOOO] [   ] [   ]. There are 8 cookies left to distribute to Chas and Delia. I cannot give 4 to each since then each would not have a different number of cookies. I cannot divide them into 3 and 5 since Alvin has 3. I could divide them into 2 and 6. The result is [OOO] [OOOO] [OO] [OOOOOO]. I now know one way to distribute the cookies: 3 to Alvin (A), 4 to Bret (B), 2 to Chas (C), and 6 to Delia (D). In fact, I know of another way because, once I give 3 to Alvin and 4 to Bret, I could also give 1 to Chas and 7 to Delia.

Could I also give 0 to Chas and 8 to Delia? When I count distributions, should I count arrangements where one of the children gets no cookies? Since there is nothing in the statement of the problem that says that each child has to get at least one cookie, I will continue under the assumption that some distributions involve some children with no cookies.

Thus, I have three different distributions at this point. Can I get any others by examining the three I already have? The first one I found was A = 3, B = 4, C = 2, and D = 6. What if I rearrange the distribution? Suppose Alvin and Bret trade and Chas and Delia trade so A = 4, B = 3, C = 6, D = 2. Is this a new distribution that should be counted? I have to decide what should be counted before I can answer how many ways the cookies can be distributed. It would be easier not to count rearrangements as new distributions, because there would be fewer to count. Since the problem statement refers to "four children" but provides no indication that the children are considered to be identifiable by name, there is some rationale for not counting rearrangements. For this reason, I will not count rearrangements.

However, if I am not going to count rearrangements as "different" distributions, I need to find some way to avoid counting them. I think the easiest way would be to arrange the groups of cookies so that they increase in number from left to right. Therefore, I will represent the distributions I have now as: [0, 3, 4, 8] and [1, 3, 4, 7] and [2, 3, 4, 6]. At this point, I believe I understand the problem, have found a partial solution, and have found a way to represent solutions. I should be ready for a more direct approach to a more complete solution.

ATTACK thoughts. Now, I will try to make an organized list that represents each of the distributions. I have one that starts with a "0," one that starts with a "1," and one that starts with a "2." I will try to find as many others as I can that start with a "0," then go on to "1," then "2" and "3" and beyond, if possible. Let's start with [0, 3, 4, 8]. I know that I can replace the 4 and 8 with numbers whose sum is 12 but that are distinct from one another and from 0 and 3. So the 4 and 8 can be replaced by 5 and 7 or by 2 and 10 or by 1 and 11. I now have the four solutions containing 0 and 3. They are [0, 3, 4, 8] and [0, 3, 5, 7] and [0, 3, 2, 12] and [0, 3, 1, 11]. I notice that the last two are not in increasing order. If the last two were ordered in increasing sequence, they would be represented as [0, 2, 3, 12] and [0, 1, 3, 11]. Perhaps it would be a good idea to find all sequences that are increasing in order and that start 0, 1; then 0, 2; then 0, 3; and so on.

Starting with 0, 1, I get [0, 1, 6, 8], [0, 1, 5, 9], [0, 1, 4, 10], [0, 1, 3, 11], [0, 1, 2, 12].

Start with 0, 2 to get [0, 2, 6, 7], [0, 2, 5, 8], [0, 2, 4, 9], [0, 2, 3, 10].

Start with 0, 3 to get [0, 3, 5, 7], [0, 3, 4, 8].

Start with 0, 4 to get [0, 4, 5, 6].

There will be none that start with 0, 5 because the sequence with the smallest sum is [0, 5, 6, 7] and that sum is greater than 15. At this point, I have five distributions that start with 0, 1; plus four distributions that start with 0, 2; plus two distributions that start with 0, 3; plus one distribution that starts with 0, 4; for a total of twelve distributions that start with 0.

Now I will construct all distributions that start with 1. I will make ordered quadruples that start with 1, that increase from entry to entry, and whose sum is 15. I will start with those that begin 1, 2 and then go on to 1, 3 and 1, 4 and continue.

Start with 1, 2 to get [1, 2, 5, 7], [1, 2, 4, 8], [1, 2, 3, 9].

Start with 1, 3 to get [1, 3, 5, 6], [1, 3, 4, 7].

Start with 1, 4, and you get no solutions because the last two entries would have to sum to 10 but be distinct and greater than 4.

At this point, I have three distributions that start with 1, 2 and two distributions that start with 1, 3—for a total of five distributions that start with 1.

Now, to construct all distributions that start with 2, there is [2, 3, 4, 6] but no other. Perhaps this means that there are no distributions that start with 3. The sequence with the smallest sum would be [3, 4, 5, 6], and that sum exceeds 15. So I have a total of twelve distributions that start with 0, five distributions that start with 1, and one distribution that starts with 2, for a grand total of eighteen distributions.

SUMMARY thoughts. I will try to justify my solution, eighteen distributions, and at the same time try to organize the distributions so that it is obvious that all possible distributions are counted and none are double-counted. It appears that the problem is equivalent to finding all sequences with four entries so that the sum of the entries is 15 and the entries are in increasing order. I constructed the following list.

[0, 1, 2, 12]	[0, 2, 3, 10]	[0, 3, 4, 8]	[0, 4, 5, 6]
[0, 1, 3, 11]	[0, 2, 4, 9]	[0, 3, 5, 7]	
[0, 1, 4, 10]	[0, 2, 5, 8]		
[0, 1, 5, 9]	[0, 2, 6, 7]		
[0, 1, 6, 8]			

[1, 2, 3, 9]	[1, 3, 4, 7]
[1, 2, 4, 8]	[1, 3, 5, 6]
[1, 2, 5, 7]	

[2, 3, 4, 6]

The patterns in the list make it unlikely that any distribution was missed or double-counted.

While working on this problem I was never really "stuck" even though I worked for some time before I began to organize things sufficiently so that I could see a way to count all of the sequences. I had to first determine precisely what was to be counted as a distribution. I also had to represent the distributions in such a way that I could distinguish them. Representing them as increasing sequences was a breakthrough. Even when I began making an organized list of sequences, I did not fully realize the power of thinking of the problem in that way. If I had realized that, I probably would have just made the list that appears above. So, while working on this problem, I learned the power of representation and organization.

Attitudes and Beliefs That Will Help to Improve Problem Solving

The most important thing is to get a lot of practice solving a lot of problems. A proper attitude is also very important. Although a proper attitude cannot guarantee success, an improper attitude does guarantee failure. The following are several important attitudes and beliefs associated with success:

- You have a problem only if you are initially "stuck." You cannot learn to solve problems without being stuck a lot. You need to develop a healthy frustration about being stuck. Healthy frustration is a motivation for learning.

- You do not learn to solve problems by watching someone else solve problems. Someone can help you by asking leading questions or by encouraging you. They cannot help by giving you answers.

- Give yourself credit for little successes and a lot of credit for larger successes. Focus on these successes rather than on the number of times you got stuck. Remember, even expert problem solvers are stuck and frustrated much more often than they achieve AHA! They have learned that experiencing all the frustration makes success that much more valued.

- Spend a lot of time with a few rich, interesting problems rather than a short time with many shallow problems. Spend time reviewing what you can get from a problem before turning your attention to another.

- When working by yourself, think out loud. Write notes to yourself. These methods help you get your thoughts out where they can be examined and corrected.

- Solve problems cooperatively with others. This experience helps because the frustration is shared and it provides you with a great opportunity to discover that everyone has problem-solving abilities.

- Don't expect to become an expert problem solver in a short time. Aim for becoming better over a longer time period. It takes time to acquire enough experience to make a difference. Even with experience, do not expect to be stuck less often. Rather expect to be able to deal with being stuck more productively.

TEACHING PROBLEM SOLVING

In a review of the literature on problem-solving research, Frank Lester, Jr. ("Musings About Mathematical Problem-Solving Research: 1970–1994." *Journal for Research in Mathematics Education* 25, no. 6, [1994]: p. 666) noted five results that may be of value to teachers of mathematics:

1. Students must solve many problems in order to improve their problem-solving ability.

2. Problem-solving ability develops slowly over a prolonged period of time.

3. In order for students to benefit from instruction, they must believe that their teacher thinks that problem solving is important.

4. Most students benefit greatly from systematically planned problem-solving instruction.

5. Teaching students about problem-solving strategies and phases of problem solving does little to improve students' ability to solve mathematics problems in general.

The implication is that students are not likely to learn to solve problems, to acquire mathematical power, unless their teacher believes it is important and conveys the sense of importance to the students. Most teachers believe that problem solving is important, but they also believe that a long list of topics and skills are important. And that list is usually ordered with the content of the local or state assessment on the top. Consequently, problem-solving activities may be postponed "until they have the background in basic skills" or "until they cover the mandated topics." As long as problem solving is regarded as an activity to be added to a crowded curriculum, rather than as an activity to be integrated into all aspects of that curriculum, it will not be possible for teachers to demonstrate to students the importance of problem solving.

A source of advice and inspiration that has been helpful and meaningful to middle school teachers in our workshops is a small book by Gary Tsuruda, *Putting It Together—Middle School Math in Transition* (Portsmouth, N.H.: Heinemann, 1994). In the book, he describes what motivated him to change his approach to teaching and to incorporate problem solving into his curriculum. He makes the case for the importance of problem solving and tells how he was able to use problem solving in the classroom. He describes how he has used a Problem of the Day (POD) and a Problem of the Week (POW). His approach exposes students to a lot of problems over a prolonged period of time and has the potential to improve problem-solving abilities of students.

The following are some suggestions and ideas about the teaching of problem solving:

- Focus attention on the importance of the "entry" phase of problem solving. One way to do that is to present a problem to your class. Tell them that you do not expect them to find a solution—or at least not a complete solution. They are to work only at the entry phase. Give them some time to play with the problem and to reach some understanding of it. Ask them to write one paragraph or more that describes what they learned about the problem situation that may be helpful in finding a solution. You may leave the writing assignment open-ended, or you may give the students some specific writing prompts to help structure their reports. For example, you might ask what questions they asked themselves and how they answered them; you might ask them to identify what they know and how they have organized that knowledge. Evaluate their abilities to make interpretations and to represent what is known.

- Distinguish between a "problem" and an "exercise," between the activity of problem solving and the repetitive activity of exercising. Some teachers, indeed some texts, present problem solving so that it is really no different from an exercise. They suggest that you can solve all problems by following these steps: Read the problem, select a strategy from a list of strategies, carry out the strategy, and check your work. They then organize all of the problem situations by what they believe to be the most appropriate strategy, so that students can practice the strategy. We do not advise introducing students to a list of strategies and providing examples of each. We prefer to have students work on a lot of problems and in the "review" stage reflect on what worked. Over a relatively short time span, they can develop their own list of strategies.

- Help your students differentiate between problem-solving strategies and ways to organize what they know. For example, making a list, making a table, drawing a graph, sketching a picture, making a Venn diagram, and making a structured or organized list are all ways to organize what you know. Since some ways of organizing what we know are more helpful than other ways in solving a certain problem, finding multiple ways to organize information is very useful in problem solving. However, organizing what we know is not a strategy in the same sense that working backward, solving a simpler problem, solving a part of the problem, and guess-check-revise are strategies. Organizing information in a way to make it most meaningful is a skill that is useful whether you are in a problem-solving situation or not. Spend time developing ways to represent what is known meaningfully.

- One thing that participants in our workshops have a difficult time learning is how to help someone solve a problem without giving him or her the answer. Some have found it easier to deal with students' questions about a problem-solving situation when they did not know the answer. If they did not know the answer, then they would work with the students, trying to get them to recall the kinds of general questions that move the process along. If they did know the answer, they were more apt to provide too much information. When the answer was revealed or too much information was provided, then the process came to a halt. In many situations, we never did reveal the answers. We did attempt to evaluate the progress made and suggest what might be done to continue. When working with students, try to ask yourself what you would do if you did not know the answer or solution and try to help the student do that. Also be very aware of the consequences of giving too much information.

- Provide a number of "metacognition" exercises. Provide a problem situation for a pair of students. Have one student try to solve the problem by thinking aloud while the second student tries to record the thoughts. This helps to focus on the process of problem solving. Students can switch roles to get the experience of both solver and recorder. As students gain more experience, the recorder can provide some of the general prompting questions as needed to move the process along. Even asking "What are you doing (or thinking)?" and "Why are you doing it?" can be useful in helping to focus attention on the process.

- Offer opportunities for students to work in groups to solve problems since that helps alleviate some of the frustration. However, try to find ways to get individual accountability. One way to do that is to start with a problem that is substantial enough that it will take some real effort. Present the problem to a class and have all students work on it individually, trying to accomplish as much of the entry phase as they can. After a predetermined amount of time, give them a few minutes to write what they understand of the problem. Then put them into groups to share what understandings they have. You can allow the groups to continue into the attack stage or return to working as individuals. With a good problem, you can go back and forth between group work and individual work several times. You can finish with each person assigned to write up a report of the process as he or she understands it. This provides all students with a work opportunity (they cannot just watch someone else work) and also with an opportunity to share and compare thinking. We have also used this combination of individual and group work in test situations successfully.

- Recognize and value "rich" problems. Rich problems will give your students an experience at each of the phases: entry, attack, and review. Rich problems usually have several different solutions. Rich problems usually come in bunches in the sense that if you solve a rich problem, you are almost immediately aware of another rich problem related to it. Rich problems are those from which we can learn the most about problem solving, about mathematics, and about ourselves. The rich problems of greatest value are those that can be used to introduce students to some of the big mathematical ideas in your curriculum.

- Acknowledge and deal openly with frustration. Students who believe that a good teacher will never allow them to become frustrated need to have that belief redirected. They need to learn that frustration is a motivation for change and for learning. They need to learn to deal with frustration instead of avoiding it. And they need help from teachers in growing to accept frustration as a normal (and honorable) state. They need the assurance that frustration is not a sign of stupidity but an opportunity for growth.

- Acknowledge and deal directly with resistance. It is not just students who may resist attempts to involve them in problem-solving activities. Some parents and administrators have the opinion that it is a teacher's job to "teach"—and that means to provide answers rather than to pose problems. They feel that too much responsibility is being placed on students to "learn for themselves." Teachers from our workshops have sent home newsletters describing what they planned to do and provided a rationale for doing it. Most were very successful with almost all parents in getting them to see the value of problem solving. They were also able to help parents understand that teachers who are teaching problem solving are not abandoning teaching and leaving students to fend for themselves. Rather, the act of teaching is helping students to find a solution that is meaningful to them instead of providing an answer that is meaningful to the teacher.

SUMMARY AND CONCLUSION

Teaching is a problem-solving behavior. The act of teaching involves an *entry* phase in which we seek to understand the circumstances of the instruction. It involves an *attack* phase in which the students are engaged in a planned activity. And it involves a *review* phase in which the teacher, as a reflective practitioner, consolidates the experience prior to beginning another cycle. Teaching is not an algorithmic behavior. There are no magic formulas, but there are general techniques that work in some circumstances. It is a continual reassessment of "What do I know about the subject, about my students, about myself?" It is a continual reassessment of "What do I want to accomplish?"

Teaching is an activity that is frequently frustrating and is punctuated with many episodes of being *stuck*. One of the fundamental frustrations occurs because teachers, by their nature, are caring individuals who naturally want to help students. Their desire to help, and the tendency of most students to accept that help, can create a relationship of dependence and leave students in a state of "learned helplessness." The responsible teacher knows that the goal of education is to produce people who become increasingly more independent of teachers. The goal of an education is to "learn how to learn." The value of accepting a problem-solving focus for the mathematics classroom is that it encourages students and teachers to accept responsibility for trying to accomplish the larger goals of education.

1
Menu for Appetizer Buffet

A LIGHT BUFFET OF APPETIZERS
TO WHET YOUR MATHEMATICAL APPETITE

1.1. Hot Dog Weight. Danny, Pat, and Kelly each tried to estimate the weight of a giant hot dog at their school fair. Danny's estimate: 59 pounds. Pat's estimate: 94 pounds. Kelly's estimate: 121 pounds. One estimate was off by 16 pounds, another by 19 pounds, and another by 43 pounds. How much did the hot dog weigh? [Feb. 1995 (4)]

1.2. Egg Timer. If you have only a 7-minute timer and an 11-minute timer, how could you time the cooking of an egg for 15 minutes? [Oct. 1998 (3)]

1.3. Special Number. Describe the number 8549176320 so that you have distinguished it from all other ten-digit numbers. [Dec. 1994 (6)]

1.4. Average to 1000. Find the average of the 1000 whole numbers from 1 to 1000, inclusive. [Oct. 1996 (1)]

1.5. Adding Red Marbles. Could you ever add enough red marbles to a container of sixteen marbles (in which the ratio of red marbles to green marbles is 3 to 5) so that the probability of drawing a green marble at random is 1? Explain fully. [Sept. 1994 (4)]

1.6. Painting Pair. Trina and Mariel were paid $60 to paint the garage. Mariel started at 8:00 A.M., and Trina did not arrive until 10:00 A.M. The work was completed at 2:00 P.M. What is Mariel's fair share of the earnings? [May 1998 (5)]

1.7. Notebook Cost. John and his twin brothers each bought three identical notebooks when school started in August. John paid for the notebooks with a $10 bill. He received less than $0.50 in change, all in nickels. No sales tax was charged. What was the price paid for one notebook? [Feb. 1995 (1)]

1.8. True or False. True or false? An average person can—

    (*a*) plant a tulip bulb in one cubic centimeter of dirt;
    (*b*) run one kilometer in a minute;
    (*c*) make ice cubes in a freezer set at 10 degrees Celsius;
    (*d*) buy a meter of milk at the grocery store. [Apr. 1995 (3)]

SOLUTIONS TO PROBLEMS
IN THE APPETIZER BUFFET

1.1. Hot Dog Weight. 78 pounds. Construct a table. The only number appearing in all three rows is 78.

	−16	−19	−43	+16	+19	+43
59	43	40	16	75	78	102
94	78	75	51	110	113	137
121	105	102	78	137	140	164

1.2. Egg Timer. Start both timers. When the 7-minute timer runs out, put the eggs on to cook. When the 11-minute timer runs out, turn it over and let it run through again. After it has run out, take the eggs off the heat. You have timed 2(11) − 7 = 15 minutes.

1.3. Special Number. The number uses each digit exactly once, and the order of the digits is alphabetical. Eight < five < four < nine ... since "e" precedes "fi," which precedes "fo," and so on.

1.4. Average to 1000. 500.5. The first 500 numbers balance the last 500 numbers when 500.5 is the center. Think of pairing the numbers 1 and 1000, 2 and 999, 3 and 998, and so on, where 500.5 is the average of each pair.

1.5. Adding Red Marbles. No. As long as the container has some green marbles, you can never be certain of drawing a red. You could not get the probability of drawing a red marble to 1 without removing green marbles.

1.6. Painting Pair. Mariel gets $36 and Trina gets $24. Mariel worked two hours alone and four hours with Trina. Trina worked four hours. They worked ten hours and earned $60, so they should get $6 an hour.

1.7. Notebook Cost. $1.10. The change could be 45, 40, 35, ... 10, 5 cents, so the cost is $9.95, $9.90, ... $9.55. The cost must be divisible by 9, the number of notebooks purchased. Only $9.90 meets that test. So each notebook cost $9.90/9 = $1.10.

1.8. True or False. All are false. You might try to determine how many cubic centimeters a flower pot might contain, how long it would take to run a kilometer, what temperature Celsius is needed to freeze water, and what metric unit you would buy milk in.

2

Menu for Dessert Buffet I

A BUFFET OF DELECTABLE MORSELS
OF RICH PROBLEMS

2.1. *Scroboscopi.* On each day of its life a scroboscopus squares its number of legs. For example, if a scroboscopus has 2 legs on its first day of life, it would have 4 legs on its second day, and 16 legs on its third day. Tom bought some newborn scroboscopi. Some had 2 legs, some had 3 legs, and some had 5 legs. The total leg count was 58 legs. The next day the leg count was 164. The following day, the leg count was 1976. How many scroboscopi did Tom buy? [Oct. 1995 (15)]

2.2. *Gym Time.* In a 36-minute gym period, 24 students want to play basketball. If only 10 can play at the same time and if each student plays the same amount of time, how many minutes does each student play? [Oct. 1996 (9)]

2.3. *Mint, Chips and Nosh.* Yolanda's Yummy Ice Cream Shoppe was giving away samples of new flavors: yum yum mint, choco chips, and nutty nosh. Forty-five people had mint, 56 had nosh, and 63 had chips. Eighteen people sampled both mint and nosh, 26 tasted chips and nosh, and 20 had chips and mint. Eight very brave people sampled all three. How many people tried the new flavors? [Nov. 1996 (5)]

2.4. *Missing Data.* The mean of a thirteen-item data set is 320. Eleven of the items are 300, 320, 199, 175, 325, 520, 156, 225, 326, 421, and 504. The median is 325. Find the remaining two items in the set if it is known that these items have the greatest possible difference. [Dec. 1996 (6)]

2.5. *Share the Estate.* An estate valued a $62 000 is left by will as follows: To each of two grandchildren a certain sum, to the son twice as much as the two grandchildren together, and to the widow $2 000 more than to the son and grandchildren together. How much goes to each? [Oct. 1997 (7)]

2.6. *Partition a Square.* A square measuring 10 centimeters on a side is divided by vertical and horizontal line segments into rectangles with areas of 12, 18, 28, and 42 square centimeters. Where should the vertical and horizontal lines be located? Draw a diagram. [Sept. 1998 (13)]

SOLUTIONS TO DESSERT BUFFET I

2.1. *Scroboscopi.* Twenty-three scroboscopi. One solution can be obtained by setting up three equations. Let a represent the number with 2 legs, b the number with 3 legs, and c the number with 5 legs. Then $58 = 2a + 3b + 5c$; $164 = 4a + 9b + 25c$; $1976 = 16a + 81b + 625c$. The solution is: $a = 15$, $b = 6$, $c = 2$.

Another approach is organized trial and error. Begin by noting that there can be no more than three with 5 legs, because $4 \times 625 > 1976$. If $c = 3$, then $16a + 81b = 101$. But there are no such whole numbers a and b. If $c = 2$, then $16a + 81b = 726$. So $b < 9$. If $b = 8$ then $16a = 78$ and there is no such a. If $b = 7$, then $16a = 159$ and there is no such a. If $b = 6$, then $16a = 240$ and $a = 15$. Thus $c = 2$, $b = 6$, and $a = 15$ satisfies the third day equation. Check to see that it satisfies the first- and second-day equations as well.

2.2. *Gym Time.* 15 minutes. Ten students playing at a time for 36 minutes represents 360 minutes of student playing time to be split among the 24 students. Since $360/24 = 15$, each student has fifteen minutes of playing time.

2.3. *Mint, Chips, and Nosh.* 108 people. A Venn diagram helps (see margin). Let M and M′ denote those who sampled mint and those who did not, respectively. Similarly let N and N′ and C and C′ denote those who sampled nosh and chips or did not. We know the number of people who had mints, nosh, and chips (MNC) = 8. Since (MN) = 18, it must be that (MNC′) = 18 − 8 = 10. Since (MC) = 20, it must be that (MCN′) = 20 − 8 = 12. Since (NC) = 26, it must be that (NCM′) = 18. Since (M) = 45, it must be that (MN′C′) = 45 − 12 − 8 − 10 = 15. Since (C) = 63, it must be that (CN′M′) = 63 − 12 − 8 − 18 = 25. Since (N) = 56, it must be that (NM′C′) = 56 − 10 − 8 − 18 = 20. So the grand total is 8 + 18 + 10 + 12 + 15 + 25 + 20 = 108.

2.4. *Missing Data.* 325 and 364. If the mean of a thirteen-item data set is 320, then the sum of the items in the set is $13(320) = 4160$. The sum of the eleven given items is 3471, so the sum of the remaining two items is 689. Because the median is 325, we know that the seventh item from the bottom is 325. Of the eleven given items, six are less than 325 and one is equal to 325. Therefore, the missing items are each ≥ 325. If they must add to 689 and if each must be ≥ 325 and have the largest possible difference, then the numbers must be 325 and 364.

2.5. *Share the Estate.* The grandchildren each get $5 000, the son gets $20 000, and the widow gets $32 000. Except for the $2 000, twelve equal parts are needed. The grandchildren each get one part, the son gets four parts (two times the grandchildren together), and the widow gets six parts (the son's part plus the grandchildren). Divide $60 000 by 12 to get the value of each part, which is $5 000.

2.6. *Partition a Square.* Divide the length or width into 6 cm and 4 cm and the other dimension into 3 cm and 7 cm. Then there will be a 3 × 6 rectangle, a 3 × 4 rectangle, a 7 × 6 rectangle, and a 4 × 7 rectangle. (See margin.)

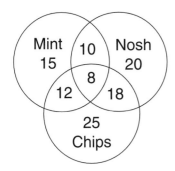

Figure for solution 2.3

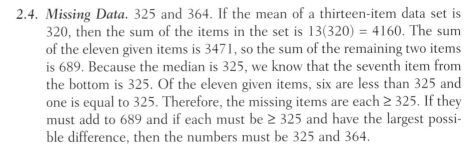

Figure for solution 2.6

3

Menu for Dessert Buffet II

DELECTABLE MORSELS OF JUST DESSERTS

3.1. *TV or Not TV.* Two banks are offering car loans. You wish to borrow $5000. The fixed payments for each loan are $100 per month. Bank A charges 1 percent interest per month on the unpaid balance, the money you still owe the bank. Bank B charges 1.5 percent interest per month on the unpaid balance and will throw in a television set valued at $1000. If you could use the TV set, which loan would you pick? [Feb. 1997 (6)]

3.2. *Chairs and Stools.* The Seats 'R' Us factory produces chairs with four legs and stools with three legs. The style of seats and legs are all the same. This year, if all the leftover legs were put with the leftover seats to make chairs, one seat would be left over without any legs. However, if all the leftover legs were put on the leftover seats to make stools, one leg would be left. How many more legs than seats are in the leftover inventory? [Sept. 1997 (5)]

3.3. *Ox and Sheep.* Two drovers hired a pasture together for $7. The first put in 4 oxen and the second, thirty sheep. What should each pay if one ox eats as much as ten sheep? [Oct. 1997 (8)]

3.4. *Squares Galore.* Find the area of the rectangle in the margin. Everything that looks like a square is a square, and the smallest square has an area of 1. [Nov. 1998 (13)]

3.5. *Connie's Boats.* Connie had two boats. She sold both boats for $6000 each. On one boat, she made a profit of 20 percent, and on the other boat, she lost 20 percent. Did Connie make money, lose money, or break even? [Nov. 1998 (12)]

3.6. *Birthday Lunch.* Ann, Barry, Charlene, Doug, Evelyn, and Fred went to lunch to celebrate Evelyn's and Fred's birthdays. Each person's meal cost the same amount. Evelyn and Fred were being treated for their meals, but each was to chip in equally for his or her share of the other's lunch. If the bill came to $54, how much should each person pay? [Jan./Feb. 1996, p. 676; Sept./Oct. 1996, pp. 40–42]

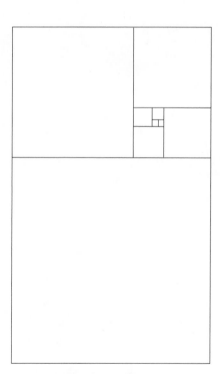

Figure for problem 3.4

17

SOLUTIONS TO PROBLEMS IN
DESSERT BUFFET II

3.1. TV or Not TV. If you choose Bank A, your first balance will be $5000, and each succeeding monthly balance is B(1.01) − 100, where B is the previous balance. It will take 71 months to pay this loan. You will have 70 payments of $100 plus a final payment on the remaining principle and interest, bringing the total payment to about $7066. For Bank B, the "1.01" must be replaced by "1.015." This loan will take 95 payments, totaling over $9400. The B loan will take over two years more to pay and cost over $2300 more in payments. You have to decide if the $1000 TV set is worth the extra cost. The calculation can be done step by step on a calculator; it can be done with a brief program on a programmable calculator; or a spreadsheet program can be used.

3.2. Chairs and Stools. Eleven. We know the difference between legs and seats in the inventory. Use the information to make a table.

	Chairs		Stools	
Pieces made (n)	Legs ($4n$)	Seats ($1+ n$)	Legs ($3n + 1$)	Seats (n)
1	4	2	4	1
2	8	3	7	2
3	12	4	10	3
4	16	5	13	4
5	20	6	16	5
6	24	7	19	6

Notice that if four chairs are made, sixteen legs and five seats would be required. If five stools were made, again sixteen legs and five seats would be required. So 16 − 5 = 11 more legs than seats in the leftover inventory.

3.3. Ox and Sheep. The oxen's cost is $4 and the sheep's $3. Since one ox eats as much as ten sheep, the four oxen would eat as much as forty sheep. So the ratio of what is eaten is 40:30, oxen to sheep.

3.4. Squares Galore. 1870 square units. The areas of squares in increasing sequence are given by the Fibonacchi sequence: 1, 1, 2, 3, 5, 8, 13, 21, 34, … The rectangle is 34 × 55.

3.5. Connie's Boats. Connie lost money. She bought one boat for $5 000 and sold it for $6 000 and made a $1 000 profit, which was $1 000/$5 000 = 20 percent profit. She bought the other for $7 500 and sold it for $6 000, which was $1 500/$7 500 = 20 percent loss. She paid $7 500 + $5 000 = $12 500 for the boats and sold them for $12 000. So she lost $500.

3.6. Birthday Lunch. $12.60 or $1.80. There were six lunches at $9 each. Fred's share is one-fifth of Evelyn's lunch, which is (1/5)($9) = $1.80. Evelyn's share is one-fifth of Fred's lunch, and that is also $1.80. Each of the others pays for his or her own lunch plus one-fifth of Fred's and one-fifth of Evelyn's, for a total of $9 + (2/5)($9) = (7)($1.80) = $12.60.

4

Food for Thought

4.1. *Mathematical Chairs.* Ten stools are arranged in a row in Jerry's Diner. Xavier, Yvonne, and Zabato go in each day and always sit so that at least one stool is between any two people. In how many different ways can the ten stools be occupied, discounting which of the three persons sit on any stool? [Mar./Apr. 1996, p. 765; Nov./Dec. 1996, pp. 93–95]

4.2. *Leading by a Head.* A six-foot-tall person walks around the earth. How much further does the person's head travel than the person's feet? [May 1997, p. 423]

4.3. *A Fraction Equivalent to a Third.* By using all of the digits 0 through 9 exactly once, make a fraction equivalent to one-third. (A figure suggests that five digits will be in the numerator and five in the denominator.) [Mar./Apr. 1997, p. 353; Nov./Dec. 1997, pp. 213–14]

4.4. *The Farmer in the Dilemma.* Once upon a time in the land of giants and beanstalks, there lived a farmer and his wife. One day the honorable tax collector came to the village and told the people that he was returning $100 of the tax moneys to them on the condition that it be used to purchase exactly 100 farm animals. The farmer was chosen by the villagers to go into town and buy some cattle, some sheep, and some horses. He was to spend the entire $100 on animals—no popcorn, jelly beans, or anything else. When he arrived in town, he discovered that sheep cost $0.50 each, cattle cost $1.00 each, and horses cost $10 each. Your job is to provide a solution to this farmer's dilemma. [Feb. 1997, p. 251; Oct. 1997, pp. 132–35]

4.5. *Nails and Pails.* A pail containing 40 nails weighs 175 grams. The same pail with 20 nails weighs 95 grams. How much does the pail weigh? How much does each nail weigh? [Jan. 1997, p. 172; Sept. 1997, p. 46–47]

4.6. *The Balance Scale.* You are given a balance scale, a lump of clay, a 50-gram weight, and a 20-gram weight. Describe how you can use these materials to produce a 15-gram lump of clay. [May 1996, p. 850; Jan. 1997, 170–71]

4.7. *Music Counts.* Your Mobile Music DJ business has six rap, ten rock, six alternative, eight "oldies," and five country CD singles. How many different ten-song sets can the DJ play at a party if she plays exactly three rap singles and four rock singles in each set? [Feb. 1998, p. 357]

4.8. *Calling All Contractors.* Iggy needs a home. He is John's pet iguana. Iggy is very active and needs plenty of room to roam. John has 60 feet of fencing that is flexible and will bend. With the available fencing, John wants to design an enclosure that will maximize Iggy's living space. Can you help him? [Mar./Apr. 1998, p. 392]

SOLUTIONS TO FOOD FOR THOUGHT

4.1. *Mathematical Chairs.* Fifty-six ways. If the first person sits on the first stool and the second person sits on the third stool, then the third person can sit anywhere from the fifth to the tenth seat—six possible seats. If the second person sits on the fourth stool, then the third person has five possible seats. Continuing, the number of possibilities when the first person occupies the first stool is $6 + 5 + 4 + 3 + 2 + 1 = 21$. If the first person sits on the second stool, then the number of arrangements is $5 + 4 + 3 + 2 + 1 = 15$. As the first person slides down in position, the number of arrangements go through the sequence $21 + 15 + 10 + 6 + 3 + 1 = 56$.

4.2. *Leading by a Head.* Assume that the earth is spherical and has radius R feet. Then the feet will travel $2\pi R$ feet. The head will travel $2\pi(R + 6)$ feet. The difference is $2\pi(6) = 12\pi$ feet.

4.3. *A Fraction Equivalent to a Third.* There are six solutions. One can be discovered by noting that $32\,000/96\,000 = 1/3$. This uses four of the ten digits and reduces the problem to one amenable to trial and error. This leads to $32\,058/96\,174$. This strategy can also be applied with $23\,000/69\,000$ to produce $23\,058/61\,749$. The other solutions are: $16\,794/50\,382$, $17\,694/53\,082$, $20\,583/61\,749$, and $30\,582/91\,746$.

4.4. *The Farmer in the Dilemma.* There are five solutions; one for each of one, two, three, four, and five horses. If you represent the number of cattle, sheep, and horses by C, S, and H, respectively, then $C + S + H = 100$ and $0.5S + C + 10H = 100$. Substituting $C = 100 - S - H$ into the second equation yields $0.5S + (100 - S - H) + 10H = 100$, which reduces to $S = 18H$. Substituting $18H$ for S in the first equation yields $C + 19H = 100$. From this, it is clear that $H \leq 5$. When $H = 5$, you can use $C + 19H = 100$ to get $C = 5$. Then $S = 90$. When $H = 4$, you use $C + 19H = 100$ to get $C = 24$. Then, $S = 72$. Continue for $H = 3$, 2, and 1.

4.5. *Nails and Pails.* The pail weighs 15 grams, and each nail weighs 4 grams. The amount of weight lost between 40 and 20 nails is $175 - 95 = 80$ grams. So 20 nails weigh 80 grams; each nail weighs $80/20 = 4$ grams. If the pail and 20 nails, each weighing 4 grams, weigh 95 grams, then the pail weighs $95 - 20(4) = 95 - 80 = 15$.

4.6. *The Balance Scale.* One solution is to put the 50-gram weight on one side of the scale and the 20-gram weight on the other side. On the side with the 20-gram weight, put a lump of clay that you alter until the scales balance and so the clay weighs 30 grams. Then remove the weights from the scale pans. Divide the clay into the two pans until the scale balances and you have two lumps of clay, each weighing 15 grams. A second solution uses only the 20-gram weight. Place clay in one pan and the 20-gram weight in the other pan until there is balance and you have a 20-gram lump of clay. Remove the weight from the scale. Divide the 20-gram lump into the two pans until they balance; resulting in two 10-gram lumps of clay. Use the same procedure on one of the 10-gram lumps to divide it into two 5-gram lumps. Finally, combine the 10-gram lump with one of the 5-gram lumps to make a 15-gram lump.

4.7. *Music Counts.* $4\,069\,800$ ways. She must choose three from six, four from ten, and three from the 19 singles remaining in the other categories. You can choose three from six in $(6 \times 5 \times 4)/(3 \times 2 \times 1) = 20$ ways. You can choose four from ten in $(10 \times 9 \times 8 \times 7)/(4 \times 3 \times 2 \times 1) = 210$ ways. You can choose three from 19 in $(19 \times 18 \times 17)/(3 \times 2 \times 1) = 969$ ways. The total number of ways is $20 \times 210 \times 969 = 4\,069\,800$.

4.8. *Calling All Contractors.* If John makes a fence that is 15 feet square, then Iggy will have 225 square feet of space. This is the greatest area for a rectangle with a perimeter of 60 feet. If John makes a circular enclosure with a circumference of 60 feet, then the radius will be found from $C = 2\pi R$, so $R = 30/\pi$. The area of a circle with radius $30/\pi$ is $\pi(30/\pi)^2 = 900/\pi$, which is approximately 286 square feet. The areas of regular enclosures with four or more sides and having a perimeter of 60 feet will be between 225 square feet and about 286 square feet.

5

Menu for Main-Course Buffet I

A BUFFET COLLECTION OF ROBUST PROBLEMS FOR ALL APPETITES

5.1. Make 1–10. Use each of the numbers 1, 2, and 4 exactly once and any of the four basic operations to make as many numbers from 1 to 10 as you can. [May 1994 (14)]

5.2. Palindrome Bus. While driving the mathematics team to a contest, the driver noticed that the odometer on the bus read 21 912. The driver also noticed that this number was a palindrome, reading the same backward as forward. The team sponsor said, "I'll bet that it will be a long time before that happens again!" However, just 105 minutes later, the driver reported that the odometer showed another palindrome. How fast was the bus traveling during that time? [Sept. 1994 (15)]

5.3. Compare Powers. Which is greater: 2^1 or 1^2? 3^2 or 2^3? 4^3 or 3^4? 7^6 or 6^7? Predict the greater of any two positive integers that follow this pattern. [Dec. 1994 (7)]

5.4. Average Speed. The mathematics team sponsor drove 120 miles to a competition going 60 mph. On the return trip, she drove 40 mph. Write an explanation to convince a teammate that the average speed is not 50 mph. Would your explanation apply even if you did not know how long the trip was? [Feb. 1995 (7)]

5.5 Jenna's Handicap. Courtney plays Jenna in a game of one-on-one that goes to 100 points and spots her 25 points to equalize the game. Lindsay plays Courtney in a game that also goes to 100 points and spots her 20 points. Lindsay is going to play Jenna in a 100-point game. How many points should Lindsay spot Jenna so that it will be a competitive game? [Feb. 1995 (8)]

5.6. Fruit Puzzle. They say you cannot add apples and oranges. However, when you subtract one fruit from another, as shown in the puzzle in the margin, you get an interesting result. Each letter represents a different digit from 0 to 8. The digit 9 does not appear. Can you break the code? [Mar. 1995 (7)]

```
   O R A N G E
 - A P P L E
 = M E L O N
```

Figure for problem 5.6

5.7. How Many Pizzas? At Pete's Pizza Parlor, a basic pizza costs $8.75. Each topping is an extra $0.50. Possible toppings are mushrooms, pepperoni, bacon, pineapple, and extra cheese. Chris has $10.00 to buy a pizza for himself and a few friends. How many different choices of pizza does Chris have? [Mar. 1995 (15)]

5.8. Storing Paper. A company delivers 100 cartons of paper to your school. Each carton measures 40 cm × 30 cm × 25 cm. The paper is stored in a space 2 m × 1.5 m × 2.5 m. Is it possible to store all the cartons of paper in this space? Explain. [Apr. 1995 (10)]

SOLUTIONS TO MAIN-COURSE BUFFET I

5.1. *Make 1–10.* Answers vary. Some examples are $(4/2) - 1 = 1$, $(4/2) \times 1 = 2$, $4 + 1 - 2 = 3$, $4/(2 - 1) = 4$, $4 + 2 - 1 = 5$, $2 \times (4 - 1) = 6$, $1 + 2 + 4 = 7$, $4 \times 2 \times 1 = 8$, $4 \times 2 + 1 = 9$, and $2 \times (4 + 1) = 10$.

5.2. *Palindrome Bus.* About 63 mph. The next palindrome is 22 022. So, 110 miles had been traveled in 105 minutes. 110 miles/1.75 hours is about 62.857 mph.

5.3. *Compare Powers.* Except for the first two instances, the expression with the greater exponent will be the greater number. That is, $n^{n+1} > (n+1)^n$ for all positive integers, n, greater than 2.

5.4. *Average Speed.* It took two hours to get there and three hours to get back, a total of five hours. The average speed would be $240/5 = 48$ mph. Note that the rate is 48 mph for a trip of any length. You can try other lengths or use algebra.

5.5. *Jenna's Handicap.* Forty points. Lindsay spots Courtney 20 points, so Lindsay expects to make 100 while Courtney makes 80. Courtney scores at a rate of $80/100 = 4/5$ compared to Lindsay. Courtney spots Jenna 25 points, so Jenna scores at a rate of $75/100 = 3/4$ compared to Courtney. Thus, Jenna to Courtney times Courtney to Lindsay, $(3/4)(4/5) = 3/5$, is the rate of Jenna to Lindsay. So Jenna should get 40 points to make the game with Lindsay even.

5.6. *Fruit Puzzle.* O = 1; R = 2; A = 6; N = 0; G = 4; E = 8; P = 7; L = 3; M = 5.

5.7. *How Many Pizzas?* Sixteen. With $10, Chris could buy a pizza with no, one, or two extra toppings. There is only one way to get a pizza with no extra toppings. There are five ways to get one extra topping; ten ways to get two extra toppings.

5.8. *Storing Paper.* Yes. The volume of each carton is 0.03 cubic meters. The volume of storage space is 7.5 cubic meters. There is ample room for 100 cartons. Students should consider different ways of stacking the cartons in the room.

6

Menu for Main-Course Buffet II

A BUFFET OF MENTAL FOOD
TO SUSTAIN HEARTY THOUGHT

6.1. *Arranging Operations.* Each of the three operation signs (+, –, and ×) is used exactly once in one of the blanks in the expression 5 __ 4 __ 6 __ 3. How many different values result from such arrangements? What are they? [Sept. 1995 (9)]

6.2. *Difference of Squares.* Two positive numbers are such that their difference is 6 and the difference of their squares is 48. What is their sum? [Sept. 1995 (12)]

6.3. *On and Off.* A train goes one-way to twenty stations and, at each station, picks up a group of people, one of which will get off at each of the remaining stations. For example, at station 5, the train picks up a group of people. One person of the group will get off at station 6, another at station 7, and so forth. At station 20, the last person of that group gets off. (*a*) What is the total number of people that get on the train? (*b*) What is the greatest number of people that will be on the train at any station? (You may assume that people getting on and off at a single station do so at the same instant.) [Oct. 1995 (11)]

6.4. *Ball, Book, and Mug.* A ball and a book cost $3.50; a book and a mug cost $6.50; a mug and a ball cost $6.00. Find the price of a ball, of a book, and of a mug. [Feb. 1996 (11)]

6.5. *Novel Operator.* If A ✿ B means (A + B)/2, then what is the value of (3 ✿ 5) ✿ 8? [Feb. 1996 (15)]

6.6. *Equivalent Fractions.* The fractions 3/6, 7/14, and 29/58 are equivalent, since each is another name for 1/2. As a set, these fractions use all the digits from 1 through 9 once and only once. Find another set of fractions similar to this set, using 3/6 and 9/18 as the first two fractions of the set. [Mar. 1996 (14)]

6.7. *Missing Score.* On five tests—on which scores could range anywhere from 0 to 100, inclusive—Johnny had an average of exactly 88. Find the lowest score Johnny could have received on one test. [Oct. 1996 (12)]

6.8. *Idle Days.* A man was hired to work for $3 a day, but on condition that for every day he was idle, he should forfeit $1.50. At the end of twenty days, he received $33. How many days had he been idle? [Oct. 1997 (15)]

SOLUTIONS TO MAIN-COURSE BUFFET II

6.1. *Arranging Operations.* There are six ways to arrange the operation signs. The numbers resulting are −9, 26, 19, −16, 23, and 17. Make sure you use the correct order of operations.

6.2. *Difference of Squares.* The sum is 8. The most immediate solution comes from recognizing that $x^2 - y^2 = (x - y)(x + y)$, and so we have $48 = 6(x + y)$; so $x + y = 8$. Another approach is to build a table of squares and look at differences.

6.3. *On and Off.* (*a*) 190 people. (*b*) 100 people. Consider the table below:

Station number	1	2	3	4	5	6	7	8	9	10	...	16	17	18	19	20
Number getting on	19	18	17	16	15	14	13	12	11	10		4	3	2	1	0
Number getting off	0	1	2	3	4	5	6	7	8	9		15	16	17	18	19
Number on	19	36	51	64	75	84	91	96	99	100					19	0

The total number that get on is the sum of the first 19 whole numbers, which is 190. Group the numbers into ten pairs, so that each pair has a sum of 19. The number of people on at each stop is the sum of those who got on minus the sum of those who got off. After station 10, more get off than get on, so the maximum cannot be on the train after station 10.

6.4. *Ball, Book, and Mug.* The ball costs $1.50; the book costs $2; the mug costs $4.50. One solution is to recognize that two of each item would cost $3.50 + $6.50 + $6.00 = $16; so one of each item would cost $8. The cost of individual items can be found by subtracting the given cost of two other items from $8.

6.5. *Novel Operator.* The value is 6. (3 ✿ 5) ✿ 8 = [(3 + 5)/2] ✿ 8 = 4 ✿ 8 = (4 + 8)/2 = 6.

6.6. *Equivalent Fractions.* 3/6, 9/18, and 27/54. We have to use 2, 4, 5, and 7 as digits to form a fraction equivalent to 1/2. The only arrangement that works is 27/54.

6.7. *Missing Score.* The missing score is 40. The average of five tests is 88, and the total of the five tests is 5(88) = 440. If Johnny scored 100 on four tests, the remaining test would be 440 − 4(100) = 40.

6.8. *Idle Days.* Six idle days. If he worked all 20 days he would have earned $60 and forfeited nothing. For every day he does not work, he loses the $3 he would have made and an additional $1.50, totaling a $4.50 loss for every day not worked. He forfeited $27 (the difference between $60 and $33), which would be 27/4.5 = 6 idle days.

7

Menu for Smorgasbord I

APPETIZERS—*A light fare of selected starters*

7.1. *People on the Bus.* Some people got on a bus. At the first stop, 2/5 of the people got off and 3/5 of the original number got on. At the second stop, 1/2 of the people got off and 1/3 of the number that were left on the bus got on. At the last stop, 3/4 of the people got off, leaving five people on the bus. How many people were on the bus before the bus reached the first stop? [Oct. 1995 (1)]

7.2. *Reds and Greens.* How many marbles should you add to a container of 16 marbles (in which the ratio of red marbles to green marbles is 3 to 5) so that the probability of drawing a green marble at random is 1/2? [Sept. 1994 (3)]

MAIN COURSES—*A smorgasbord of hearty fare*

7.3. *Knife, Fork, and Spoon.* On a balance scale, a knife balances a spoon and a fork, five spoons balance a knife and a fork, and a plate balances a knife and a spoon. You can represent this symbolically by $k = s + f$, $5s = k + f$, and $p = k + s$. If a fork weighs four ounces, how much does a plate weigh? [Mar. 1996 (7)]

7.4. *Boxes inside Boxes.* Inside a large box are six smaller boxes. Some of these smaller boxes are empty, and others have six very small empty boxes inside them. Thirty-one boxes in all are present. How many of these thirty-one boxes are empty? [Jan. 1998 (11)]

7.5. *Discount This.* A local building supply store claims that its "everyday low price" is 70 percent off retail. During a special sale, the store advertises that its prices will be 75 percent off retail. What percent discount is the store offering off its regular low prices? [May 1998 (12)]

DESSERTS—*A fine selection of finishing dishes*

7.6. *Count Off.* At a school, the children lined up in a long row and counted off. A hat was given to the sixteenth child in line and every sixteenth child after that. A noisemaker was given to the twenty-fourth child and to every twenty-fourth child after that. What were the positions in line of the first three children who received both a hat and a noisemaker? [Feb. 1996 (18)]

7.7. *Test Scoring.* A mathematics contest consisted of twenty problems. Three points were given for each correct answer, one point was subtracted for each incorrect answer, and no points were awarded for items that were skipped. Your team scored 48 points. How many answers were incorrect and how many items were not answered? [Dec. 1996 (5)]

SOLUTIONS TO SMORGASBORD I

7.1. *People on the Bus.* Twenty-five people. Twelve-year-old Andrew Glendening's solution: At the first stop, we lose 2/5 of the people, leaving 3/5 of the original number on the bus. We pick up an additional 3/5, giving us 6/5 of the original number. At the second stop, we lose half of the 6/5, leaving 3/5. We pick up 1/3 of 3/5 = 1/5. The five remaining people represent 1/5 of the original number.

7.2. *Reds and Greens.* Four red marbles. You need to add enough red marbles so the ratio of red to green is 1 to 1. You start with 6 red and 10 green, so you must add 4 red to get to 10 of each color.

7.3. *Knife, Fork, and Spoon.* Eight ounces. From the first two statements, we have $s + s + s + s + s = s + f + f$. By removing a spoon from each side, we get $s + s + s + s = f + f$. By taking half of each side, we get $s + s = f$. Since a fork weighs exactly 4 ounces, a spoon must weigh 2 ounces. Since $k = s + f$, a knife must weigh $2 + 4 = 6$ ounces. Finally, since $p = k + s = 6 + 2 = 8$, a plate must weigh 8 ounces.

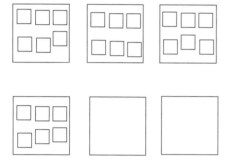

Figure for solution 7.4

7.4. *Boxes Inside Boxes.* Twenty-six boxes are empty. We are given one large box and six smaller boxes, for a total of seven boxes. We still need twenty-four more boxes to have thirty-one boxes all together. Therefore, of the second-sized boxes, four will contain six smaller boxes. All of those twenty-four boxes will be empty, plus two of the middle-sized boxes, making a total of twenty-six empty boxes. (See figure on the left.)

7.5. *Discount This.* 16 2/3 percent. The store's prices are 30 percent of the retail price. The sale price is 25 percent of the retail price. So the reduction is 5 percent off the base of 30 percent: 5/30 = 16 2/3 percent.

7.6. *Count Off.* Numbers 48, 96, and 144 are the first three common multiples of 16 and 24.

7.7. *Test Scoring.* No answers incorrect and four unanswered OR three answers incorrect and no unanswered items. Since 48/3 = 16, your team had at least sixteen correct out of twenty. In fact, 48 points could be obtained by getting sixteen correct and omitting the other four items. If there were seventeen correct answers, then there would have to be three incorrect and no unanswered items to obtain 48 points. The team could not have gotten eighteen or more correct and still have scored 48 points, because too few items are left to get wrong and bring the score down to 48.

8

Menu for Smorgasbord II

APPETIZERS—*Light delights*

8.1. ***Nine Coins.*** You have just received nine rare coins. You get an anonymous tip that one of them is counterfeit. The real coins weigh 10 grams each. The counterfeit looks just like the genuine coins, but it weighs less. The only scale you have is an old balance scale with two pans, one on each side of a balance point. What is the fewest number of weighings you can make and identify the counterfeit? Would your answer change if you knew that the counterfeit coin was heavier than the genuine coins? Would your answer change if you knew only that the counterfeit coin was of a different weight than the genuine coins? [Nov. 1998 (1)]

8.2. ***Missing Weight.*** Find the missing weight on the scale. One block of weight A and one block of weight B weigh 90 kg. Two blocks of weight A and one block of weight B weigh 115 kg. How much do three blocks of weight A and one block of weight B weigh? [Oct. 1994 (6)]

MAIN COURSES—*Hearty fare*

8.3. ***Tiling, Tiling.*** A rectangular tile measures 2 inches × 3 inches. What is the least number of tiles that are needed to completely cover a square region two feet on a side? [Sept. 1996 (13)]

8.4. ***Fill the Tub.*** A bathtub can be filled by the hot and cold water pipes in four minutes and, when full, can be emptied by the waste pipe in five minutes. If it is half full and all three pipes are set running, when will the tub be filled? [Oct. 1997 (12)]

8.5. ***Elf on the Stairs.*** In how many different ways can an elf climb a ten-step staircase if he or she climbs only one or two steps at a time and always goes up? [Jan. 1998 (12)]

DESSERTS—*Delectables*

8.6. ***Amelia's Panda.*** Amelia was trying to win a giant stuffed panda at the fair's "bear toss." She had to hit cutout figures of ten small bears and twelve large bears to win a panda. Every small bear was worth 25 points and every large bear was worth 15 points. Amelia hit twenty-two bears for 410 points. Did she win the giant panda? Explain. [Sept. 1996 (6)]

8.7. ***Sibling Puzzler.*** If I have two more brothers than sisters and if each of my brothers also has two more brothers than sisters, how many more brothers than sisters does my oldest sister have? [Oct. 1996 (8)]

8.8. ***Granny's gift.*** On the day you were born, your generous grandmother put $1000 into a savings account for you. She told your parents that you could have the money when it had doubled. How long will it take to double your investment if you earn 10 percent compounded annually? To double a second time? [Nov. 1997 (6)]

SOLUTIONS TO SMORGASBORD II

8.1. *Nine Coins.* Two. Place three coins in one pan and three in the other. If the pans balance, the counterfeit is one of the three remaining coins. If the pans do not balance, the counterfeit is in the pan that is higher. In either instance, you have reduced the search to three coins. Place one coin in one pan and one in the other. If the pans balance, the counterfeit is the odd coin. If the pans do not balance, the counterfeit is in the higher pan. If you knew the counterfeit was heavier, two weighings would suffice. (Look in the lower pan, rather than the higher.) If you only knew that the counterfeit was a different weight, more than two weighings would be needed.

8.2. *Missing Weight.* 140 kg. The difference in weights on the first two scales is 25 kg; so this must be the weight of one A block. The third scale adds 25 kg to the second, so the weight of the third must be $115 + 25 = 140$ kg.

8.3. *Tiling, Tiling.* Ninety-six tiles. The square region is 24 inches × 24 inches, and its area is 576 square inches. Each small tile has an area of 6 square inches, so there must be at least $576/6 = 96$ tiles. Note that the tiles can be put in a 12×8 array to cover the square.

8.4. *Fill the Tub.* Ten minutes. The filler pipes can fill 1/4 of the tub in a minute. The waste pipe would empty 1/5 of the tub each minute. With both operating, the net gain per minute is $1/4 - 1/5 = 1/20$ of the tub. Therefore, the tub would take twenty minutes to fill if it started empty. Since it started half full, it should take ten minutes.

8.5. *Elf on the Stairs.* Eighty-nine ways. Consider one step. The elf has only one way to climb it. For two steps, the elf could go one step, then another step, or he or she could take two steps at once. For three steps, the elf could go one, one, one; two, one; or one, two, for a total of three ways. Continuing, you see the following table:

Number of steps	1	2	3	4	5	6	7	8	9	10
Number of ways	1	2	3	5	8	13	21	34	55	89

The result is the Fibonacci sequence. Why? Consider seven steps. The elf can climb either one step leaving six, which could be done in thirteen ways; or two steps, leaving five, which could be climbed in eight ways; so the total is $13 + 8 = 21$ ways.

8.6. *Amelia's Panda.* No. If Amelia would have hit ten small bears and twelve large bears, then she would have scored $10(25) + 12(15) = 430$ points. Since she only scored 410 points, she could not have won. She hit twenty-two bears, but she hit eight small bears and fourteen large bears.

8.7. *Sibling Puzzler.* Four. Make a table that shows the speaker has two more brothers than sisters.

Number of sisters	0	1	2	3	4	5	6
Number of brothers	2	3	4	5	6	7	8

Suppose the speaker is a girl. If she has two brothers, then each brother would have only one brother, not the two required in the problem. If she had three brothers, then each brother would have two sisters and two brothers, not the two more brothers than sisters that are required. If she has four brothers, then each brother would have three brothers and three sisters, not the two more brothers as required. The situation is similar if she has five or more brothers. So the speaker cannot be a girl. If the boy who is the speaker has two brothers, then he would have no sisters, which contradicts his having an oldest sister. If he has three brothers, then he has one sister; and each of his three brothers has three brothers and a sister. The sister has four brothers and no sisters; a difference of four.

8.8. *Granny's Gift.* The account will double in less than 8 years. At the end of year t, the amount in the account, is $1000(1.1)^t$. When $t = 8$, the value is about 2143.59, and when $t = 15$, the value is about $4174.80.

9

Menu for Problems about the Calendar

APPETIZERS—*Light fare to whet your appetite for calendar problems*

9.1. *Count Prime Dates.* A prime day has a month and day which are both prime. Thus, 5/13 is a prime date. How many prime dates are there in the year 2000? [Dec. 1998 (3)]

9.2. *Last Prime Date.* If both the month and the day are prime numbers, then consider this date to be a prime date. For example, July 19 (7/19) is a prime date. What are the first and last prime dates of the year? [Sept. 1996 (2)]

MAIN COURSE—*Mental food to sustain you for all dates*

9.3. *Product Dates.* For the year 1994, no date exists for which the product of the month and day equals the number named by the last two digits of the year. In 1993, the date 3-31-93 met the criteria. During this century, 1901–2000, which year has the most possibilities? [Apr. 1994 (16)]

9.4. *Vote Date.* National elections occur on the first Tuesday after the first Monday in November. Which dates are possible? [Oct. 1995 (9)]

9.5. *Factor Days.* For how many days during the year is the number of the day a factor of the number of the month? [Nov. 1998 (8)]

9.6. *Multiple Days.* For how many days during the year is the day a multiple of the month? [Nov. 1998 (8)]

DESSERT—*Delectable morsels of days and dates*

9.7. *Get into the Swim.* Mary and Sandy are on the same swim team for the summer. The coach has placed each team member on a different training schedule. As a result, Mary will swim laps every other day and Sandy will swim laps every three days. The schedule will begin June 5, with both girls swimming laps on that day. How many times will Mary and Sandy swim laps on the same day during the first ten weeks of the summer? Suppose that the girls decide to keep the same schedule until their winter vacation, which begins December 20. How many times will they swim together from June 5 through December 20? After the girls plan their schedules for both the summer and the fall semester, the coach is notified that the pool will be closed for lap swimming on these holidays: the Fourth of July, Labor Day, and Thanksgiving Day. How does this schedule change the girls' original plans? [May 1998, p. 483]

SOLUTIONS TO PROBLEMS ABOUT THE CALENDAR

9.1. *Count Prime Dates.* Fifty-three. Prime months are 2, 3, 5, 7, and 11; prime days are 2, 3, 5, 7, 11, 13, 17, 19, 23, 29, and 31. If all prime months had 31 days, there would be 55 prime dates. But February in 2000 has 29, and November has 30. So there are 53 prime dates.

9.2. *Last Prime Date.* 2/2 and 11/29. The smallest prime is 2, so 2/2 would be the first prime date. The largest prime less than or equal to 12 is 11. November, the eleventh month, has 30 days. The largest prime less than or equal to 30 is 29.

9.3. *Product Dates.* 1924. You want a year that is divisible by the most month-numbers (2, 3, 4, 5, 6, 7, 8, 9, 10, 11, and 12). It appears sufficient to check multiples of 12. A check reveals that there are four dates in 1996; five dates in 1984; six dates in 1972, 1960, 1948, 1936, and 1912; and seven dates in 1924.

9.4. *Vote Date.* November 2nd through the 8th. The first Monday can be on the 1st through the 7th, and Tuesday is the following day.

9.5. *Factor Days.* Thirty-five times. Consider this on a month-by-month basis.

Month Number	1	2	3	4	5	6	7	8	9	10	11	12
Number of Factors	1	2	2	3	2	4	2	4	3	4	2	6

9.6. *Multiple Days.* Ninety times. Consider this on a month-by-month basis.

Month Number	1	2	3	4	5	6	7	8	9	10	11	12
Number of Multiples	31	14	10	7	6	5	4	3	3	3	2	2

9.7. *Get into the Swim.* Ten weeks is 70 days. The girls will swim together every sixth day. That is, they will swim together on days 1, 7, 13, 19, 25, 31, 37, 43, 49, 55, 61, and 67. The list has 12 days in it. If they continue until December 20, they will swim on 26 days in June; 31 days in July, August, and October; 30 days in September and November; and 20 days in December, for a total of 199 days. Since 199 = 33 × 6 + 1, they will swim together on December 20 for the 34th time. Having the pool closed on July 4th will not effect the number of times they swim together for the first ten weeks in the summer since they swim together 12 times in 67 days. There is enough room for a break on the Fourth of July. However, the three holidays do effect the number of times they swim together by December 20. Missing three days means they will swim together only 33 times by December 20.

10
Menu for Clock Problems

APPETIZER—*Light fare to whet your digital appetite*

10.1. Fives. On a digital clock showing hours and minutes, how many different readings between noon and 6:00 P.M. contain at least two 5s? [Apr. 1994 (3)]

MAIN COURSES—*Mental food of a timely nature*

10.2. Palindrome. How many times during a day will a digital clock show a palindrome? [Jan. 1997 (9)]

10.3. Right Angle. How many times during a twenty-four-hour day would the minute and hour hands of a watch form a right angle? [Apr. 1997 (15)]

10.4. Angle at 2:48. What is the number of degrees in the measure of the obtuse angle formed by the hands of a standard clock at 2:48 P.M.? [Sept. 1997 (15)]

DESSERTS—*Delectable morsels to enrich the spirit for all times*

10.5. Digital Changes. The display on a digital clock reads 6:38. What will the clock display twenty-seven digit changes later? [Oct. 1995 (13)]

10.6. Increasing. In any twelve-hour period how many times will the digits of a digital clock be in strictly increasing order? [Dec. 1998 (7)]

10.7. Decreasing. In any twelve-hour period how many times will the digits of a digital clock be in strictly decreasing order? [Dec. 1998 (8)]

10.8. Angle at 5:20. What is the measure in degrees of the acute angle formed by the hour and minute hands of a clock at 5:20 P.M.? [Sept. 1995 (20)]

10.9. Losing Time. A clock that uniformly loses four minutes every twenty-four hours was set correctly at 6:00 A.M. on 1 January. What was the time indicated by this clock when the correct time was 12:00 NOON on 6 January of the same year? [Oct. 1996 (11)]

SOLUTIONS TO CLOCK PROBLEMS

10.1. Fives. Twenty readings. Five times before 5:00 and five times between 5:05 and 5:45; ten times from 5:50 to 5:59.

10.2. Palindrome. 114 times. For each hour, n, when $1 \leq n \leq 9$, there will be six times of the form $n{:}0n$, $n{:}1n$, $n{:}2n$, $n{:}3n$, $n{:}4n$, $n{:}5n$. This accounts for 54 palindromes. Then there are 10:01, 11:11, and 12:21, a total of three more. There are 57 in a twelve-hour period and, therefore, 114 in a day.

10.3. Right Angle. Forty-four times. Starting at 12:00 midnight, the hands would form a right angle at approximately 12:15, 12:45, 1:20, 1:40, 2:25, 3:00, 3:35, 4:05, 4:40, 5:10, 5:45, 6:15, 6:50, 7:20, 7:55, 8:25, 9:00, 9:35, 10:05, 10:40, 11:10, and 11:45. They will be repeated in the next twelve hours. You may reason that in a twelve-hour period, the minute hand will pass the hour hand eleven times and form a "straight angle" eleven times. Between passing and straight, a right angle will result; between straight and passing, another right angle will be formed. This totals twenty-two in a twelve-hour period.

10.4. Angle at 2:48. 156 degrees. We find the size of the angle by finding the measures of three pieces and then adding them. The first piece is between the minute hand at 48/60 and the top of the clock at 60/60. The distance is 12/60 = 1/5. The second piece is between the top of the clock at 0/12 and the hour hand at 2/12. The distance is 2/12 = 1/6. The third piece is between the hour hand at 2:00 and the hour hand at 48/60 = 4/5 of an hour later. This distance is (4/5)(1/12) = 1/15. We add 1/5 + 1/6 + 1/15 to get 13/30. The angle is 13/30 of 360 degrees, which is 156 degrees.

10.5. Digital Changes. The time will be 7:01. From 6:38 to 6:39, there is one digit change. From 6:39 to 6:40, there are two. From 6:40 to 6:41 to 6:42 to … 6:49, there are nine single digital changes. From 6:49 to 6:50, there are two; from 6:50 to 6:51 to 6:52 to … to 6:59, there are nine more single digital changes. There are 23 up to 6:59. From 6:59 to 7:00, there are three more, for a total of 26. From 7:00 to 7:01, there is the 27th digital change.

10.6. Increasing. Sixty-five times. Consider this list of the twenty-two times between 1:00 and 2:00:

1:23	A	B	C
1:24	1:34		
1:25	1:35	1:45	
1:26	1:36	1:46	1:56
1:27	1:37	1:47	1:57
1:28	1:38	1:48	1:58
1:29	1:39	1:49	1:59

The hours in columns A, B, and C can be replaced with "2," thus obtaining fifteen more times. The hours in columns B and C can be replaced with "3," obtaining nine more times. The hours in column C can be replaced with "4," obtaining four more times. There are no times in the 10:00 or 11:00 hour. The hours in columns A, B, and C can be replaced with "12," obtaining fifteen more times. So we have 22 + 15 + 9 + 4 + 15 = 65 times.

10.7. Decreasing. Eighty times. The first time is 2:10. Then there are 3:21, 3:20, and 3:10. Then there are 4:32, 4:31, 4:30, 4:21, 4:20, and 4:10. This begins a pattern of triangular numbers: 1, 3, 6, 10, and 15. There are ten times in the 5:00 hour and fifteen times in the 6:00 hour. Because the upper limit on minutes is 59, there are also fifteen times in the 7:00, 8:00, and 9:00 hours. So we have 1 + 3 + 6 + 10 + 4 × 15 = 80.

10.8. Angle at 5:20. Forty degrees. The number of degrees between the "4" and the "5" is 360/12 = 30 degrees. The hour hand travels 30 degrees in sixty minutes. So, in twenty minutes, the hour hand has moved 10 degrees away from the "5." Thus, the angle between the hour hand and the minute hand, which is pointing directly at the "4," is 40 degrees.

10.9. Losing Time. 11:39 A.M. At 6:00 A.M. on the following dates, these times will occur: 2 January, 5:56 A.M.; 3 January, 5:52 A.M.; 4 January, 5:48 A.M.; 5 January, 5:44 A.M.; 6 January, 5:40 A.M. If the clock loses time uniformly at a rate of four minutes in twenty-four hours, it will lose one minute in six hours, so at noon on 6 January, the clock will have lost twenty-one minutes and show 11:39 A.M.

11
Menu for Counting Problems

APPETIZERS—*Light fare of counting situations*

11.1. Different Sums. Mike and Sonja are playing a game with two numbered cubes. One of them is numbered 0, 1, 2, 3, 4, 5. The other is numbered 0, −1, −2, −3, −4, −5. How many different sums are possible when the two cubes are rolled? [Mar. 1995 (6)]

11.2. Make Change. In how many ways can you make change for a dollar using only dimes or nickels? [Apr. 1997 (3)]

MAIN COURSES—*Robust counting situations*

11.3. Toppings. A billboard advertises that a restaurant has more than 1000 varieties of pizza. If a plain pizza contains just cheese and tomato sauce, how many toppings does the restaurant have? [May 1995 (16)]

11.4. Pass the Cookies. Moe, Larry, and Curley had eight cookies to distribute among themselves. They all do not need to get the same number of cookies, but they must all get at least one cookie. In how many ways could the cookies be distributed? [Dec. 1997 (13)]

11.5. Pass Again. If Harpo joined the Three Stooges (see #11.4 above) before they figured out this problem, and he also needed to be included in the distribution of the cookies, in how many ways could the cookies be distributed? [Dec. 1997 (14)]

DESSERTS—*Count on these to finish a meal*

11.6. Four-Digit Palindromes. A palindrome is a number that reads the same from right to left as it does from left to right. For example, 363, 77, and 24642 are palindromes. How many four-digit palindromes are there? [Jan. 1996 (18)]

11.7. Fruit Juggler. A man is juggling three apples, four oranges, and seven grapefruit. He drops eight pieces of fruit. How many possible combinations of fruit remain for him to juggle? [Jan. 1998 (6)]

11.8. DJ's Choice. Your Mobile Music DJ business has six rap, ten rock, six alternative, eight "oldies," and five country CD singles. How many different ten-song sets can the DJ play at a party if she plays exactly two singles from each category? [Feb. 1998, p. 357]

SOLUTIONS TO COUNTING PROBLEMS

11.1. *Different Sums.* Eleven sums.

+	0	1	2	3	4	5
0	0	1	2	3	4	5
−1	−1	0	1	2	3	4
−2	−2	−1	0	1	2	3
−3	−3	−2	−1	0	1	2
−4	−4	−3	−2	−1	0	1
−5	−5	−4	−3	−2	−1	0

11.2. *Make Change.* Eleven ways. You can use 0, 1, 2, 3, 4, 5, 6, 7, 8, 9, or 10 dimes with the remainder being nickels.

11.3. *Toppings.* At least ten toppings. The number of choices is 2^n, where n is the number of toppings. Since $2^{10} = 1024$, there must be at least ten toppings.

11.4. *Pass the Cookies.* Twenty-one ways. Since they all need to get at least one cookie, this is the same as partitioning 8 using three addends. There are three ways to distribute them for 1–1–6, six ways to distribute them for 1–2–5, six ways for 1–3–4, three ways for 2–2–4, and three ways for 2–3–3.

11.5. *Pass Again.* Thirty-five ways. We need to partition 8 into four addends. There are four ways to get 1–1–1–5, twelve ways to get 1–1–2–4, six ways to get 1–1–3–3, twelve ways to get 1–2–2–3, and one way to get 2–2–2–2.

11.6. *Four-Digit Palindromes.* Ninety. Since the palindrome must have four digits, the first digit must be one of the nine nonzero digits. The second digit can be any one of the ten digits. The third and fourth digits are determined by the first two. So there are $9 \times 10 = 90$ ways to make a four-digit palindrome.

11.7. *Fruit Juggler.* Nineteen possibilities. He dropped eight of the fourteen pieces, leaving six pieces. Make an organized list, using the fact that the number of apples might be 3, 2, 1, or 0. If there are three apples, there can be 3, 2, 1, or 0 oranges. If there are two, one, or no apples, there can be 4, 3, 2, 1, or 0 oranges. So there are $3 \times 5 + 4 = 19$ possibilities.

11.8. *DJ's Choice.* 2 835 000. You must select 2 from 6, 2 from 10, 2 from 6, 2 from 8, and 2 from 5, where the number of ways to select 2 from n is $n(n-1)/2$. The number of selections is $15 \times 45 \times 15 \times 28 \times 10 = 2\,835\,000$, assuming the song sets are not ordered.

12

Menu for Problems on Cubes

APPETIZERS—*Light fare of items from cubes*

12.1. Surface Area. Some unit cubes are placed side by side to form rectangular prisms. Determine the surface area for prisms formed from one, two, three, four, or five cubes. Write an expression for finding the surface area of a prism with *n* cubes. [Jan. 1997 (4)]

12.2. Pack a Box. How many $2'' \times 2'' \times 2''$ cubes will fit into a $3'' \times 3'' \times 3''$ box? [Feb. 1998 (2)]

MAIN COURSES—*Mental food to satisfy hunger for cubes*

12.3. Painted Cubes. A number of unit cubes are put together in the form of a right rectangular prism. The six faces of the prism are painted. When the unit cubes are taken apart, it is found that 363 of them have no paint on them. What is the volume of the prism that was formed? [Feb. 1998 (16)]

DESSERTS—*Delectable morsels to enrich experience with cubes*

12.4. Painted Dice. You are given a $3'' \times 3'' \times 3''$ cube. The cube is painted and then cut up into twenty-seven 1-inch cubes. You place all twenty-seven cubes in a large paper sack and shake it so that the cubes are well mixed. Randomly select one cube from the bag and roll it as if it were a die. What is the probability that the top face of the cube is painted? [May 1994 (18)]

12.5. General Painted Dice. Suppose you have a $4'' \times 4'' \times 4''$ cube. The cube is painted and then cut up into sixty-four 1-inch cubes. You place all cubes in a bag, mix them well, and then draw a cube. The drawn cube is rolled like a die. What is the probability that the top face of the cube is painted? What if the original cube were $5'' \times 5'' \times 5''$? Can you generalize to an $N \times N \times N$ cube? [May 1994 (19)]

12.6. Counting Colors. Mrs. Gonzales has some white paint, some red paint, and a bunch of wooden cubes. Her class has decided to paint the cubes by making each face either solid red or solid white. One student painted his cube with all six faces white. Another painted her cube with all six faces red. Yet another painted four faces white and two red. How many cubes could be painted in this fashion so that each cube is different from the others? Two cubes are different if, no matter which way you turn them, they cannot be matched up. [Oct. 1994 (18).]

12.7. Remove Cubes. Start with a cube measuring three inches on an edge and remove a one-inch cube from the center of each face. (See figure below.) What is the surface area of the remaining structure? [Jan. 1995 (12)]

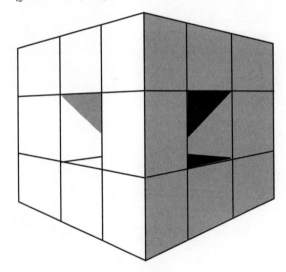

Figure for problem 12.7

12.8. One Cube to Three. A solid metal cube nine inches on a side is cut into three parts. Each part is then reshaped to form a perfect cube. One is eight inches on a side. Another is one inch on a side. What is the size of the third cube? [Jan. 1998 (8)]

SOLUTIONS TO PROBLEMS ON CUBES

12.1. *Surface Area.* $4n + 2$. For the first five prisms, the areas are 6, 10, 14, 18, and 22. Each prism has two square faces on opposite ends. Each cube contributes four square units to the surface area contained in the remaining faces; hence, $4n + 2$.

12.2. *Pack a Box.* One cube. The smaller cube is 8 cubic inches; the larger is 27 cubic inches. So the answer could be 1, 2, or 3. However, once one cube is placed in the box, no other cubes will fit.

12.3. *Painted Cubes.* 845 cubic units or 3285 cubic units. The 363 unit cubes can be arranged in the form of a right rectangular prism that has dimensions $11 \times 11 \times 3$ because the prime factorization of 363 is $3 \times 11 \times 11$. None of the faces of the 363 cubes are painted, so they must be surrounded on all sides by other cubes. The larger prism must be $13 \times 13 \times 5$, and it would have a volume of 845 cubic units. Another possible arrangement for the unpainted cubes would be $1 \times 1 \times 363$. Then the surrounding prism would be $2 \times 2 \times 365$, and its volume would be 3285 cubic units.

12.4. *Painted Dice.* 1/3. The twenty-seven cubes have a total of $27 \times 6 = 162$ faces. There are $9 \times 6 = 54$ painted faces. The probability of picking a cube and then rolling a painted face must be $56/162 = 1/3$.

12.5. *General Painted Dice.* In general, the probability is $1/N$. The total number of faces painted is $6N^2$, and the total number of faces on the small cubes is $6N^3$, so the probability of getting a painted face is $6N^2/6N^3 = 1/N$.

12.6. *Counting Colors.* Ten.

Colors	6W	5W	4W	3W	2W	1W	0W
Number of colorings	1	1	2	2	2	1	1

(Note: "6W" refers to six faces painted white)

12.7. *Remove Cubes.* 78 square inches. A cube has six faces. After the cubes are removed, each original face has eight squares that contribute to the surface area for a total of $6 \times 8 = 48$ square inches. One inside hole is on each of the original six faces, and five faces of the hole are exposed, for a total of $6 \times 5 = 30$ square inches. Add $48 + 30$ to get 78 square inches.

12.8. *One Cube to Three.* Six inches on a side. The original cube has a volume of $9 \times 9 \times 9 = 729$ cubic inches. The 8-inch cube has a volume of 512 cubic inches. The 1-inch cube has a volume of 1 cubic inch. The third cube must have $729 - 512 - 1 = 216$ cubic inches. But $6 \times 6 \times 6 = 216$.

13

Menu for Problems about Digits

APPETIZERS—*Light fare to whet your appetite for digits*

13.1. "3" in Tag. Allison is making number tags for the coat check for the school dance. She must make two sets of tags, which are numbered 1 through 100. How many times will she write the digit 3? [Mar. 1995 (4)]

13.2. Digits on Tickets. Whitney had to write numbers on tickets for the school play. The ticket numbers ranged from 1 to 325. How many digits did she write? [May 1997 (1)]

13.3 Sevens. What fraction of the numbers from 1 to 1000 have the digit 7 as one of the digits? [Oct. 1998 (1)]

MAIN COURSE—*Mental food to sustain counting skills*

13.4. The Last One. Betty wrote a list of consecutive whole numbers, starting with the number 1. She used 261 digits. What was the last number she wrote? [Jan. 1995 (8)]

13.5. Magic Triangle. Place the digits 1 through 9 in the circles on the "magic triangle" in the margin using each digit exactly once. For each side of the triangle, the sum of the two "inside numbers" subtracted from the sum of the two "end numbers" must be the same "magic number." Find the different arrangements for the magic numbers 0, 3, and 9. [Sept. 1994 (14)]

13.6. Add to Eight. In the range of numbers between 1 and 209, how many times will the digits of a number add to eight? [Dec. 1996 (16)]

DESSERTS—*Delectable morsels to enrich and satisfy*

13.7. Odd and Even. Find the percent of the whole numbers from 0 to 99 that—

(*a*) contain only even digits;
(*b*) contain only odd digits;
(*c*) contain both even and odd digits. [Apr. 1995 (20)]

13.8. Number a Cube. Number the eight corners of a cube from 1 to 8 so that the sum of the four numbers at the corners of each face is 18. [Jan. 1997 (5)]

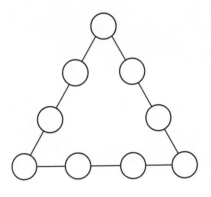

Figure for problem 13.5

SOLUTIONS TO PROBLEMS ABOUT DIGITS

13.1. *"3" in Tag.* Forty. Can be done by making a list. {3, 13, 23, 30, 31, 32, 33, 34, 35, 36, 37, 38, 39, 43, 53, 63, 73, 83, 93}. There are twenty 3s in the list, which must be duplicated.

13.2. *Digits on Tickets.* 867 digits. From 1 to 9, you use nine digits. From 10 to 99, you use (90)(2) = 180 digits. From 100 to 325, you use (226)(3) = 678 digits. In all, use 9 + 180 + 678 = 867 digits.

13.3. *Sevens.* 271/1000. Since 7 occurs nineteen times in each of 100 numbers (ten times in the seventies and nine other times), there will be 19 times 9 = 171. This does not count the 700s, which are another 100 numbers.

13.4. *The Last One.* 123. In writing 1–9, she used nine digits. In writing 10–99, she used $90 \times 2 = 180$ digits. She has left $261 - 9 - 180 = 72$ digits to use on three-digit numbers. She can write 72/3 = 24 three-digit numbers starting with 100.

13.5. *Magic Triangle.* For magic number 0, starting at the top and going clockwise, use 4, 7, 3, 6, 2, 9, 5, 1, 8. For magic number 3, starting at the top and going clockwise, use 6, 7, 5, 9, 1, 8, 3, 4, 2. For magic number 9, starting at the top and going clockwise, use 9, 1, 6, 7, 2, 4, 8, 5, 3.

13.6. *Add to Eight.* Eighteen. The list is 8, 17, 26, 35, 44, 53, 62, 71, 80, 107, 116, 125, 134, 143, 152, 161, 170, and 206. There is one in each decade except for the 90s, the 180s, and the 190s.

13.7. *Odd and Even.*

(*a*) 25 percent. Five ways to make a tens digit and five ways to make a units digit.

(*b*) 30 percent. Five ways to make numbers in each decade 0–9, 10–19, 30–39, and so forth.

(*c*) 45 percent. 100 percent – 25 percent – 30 percent = 45 percent.

13.8. *Number a Cube.* To get a sum of 18 using four addends, zero, two or four of the addends must be odd. Since the odd numbers 1, 3, 5, and 7 add to 16, we know that we cannot use zero or four of them on a face. We must put exactly two odd numbers on each face. So put 1 and 7 on opposite corners of the top face and put 3 and 5 on opposite corners of the bottom face, so that no two odd numbers are endpoints of an edge. Then numbers 4 and 6 must be put with 1 and 7, and 2 and 8 must be placed with 3 and 5. (See margin.)

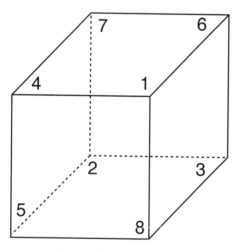

Figure for solution 13.8

14

Menu for Problems about Factors and Multiples I

APPETIZERS—*Light fare to multiply your appetite*

14.1. Mystery Number. The numbers 3, 5, and 12 are three of the twelve factors of a mystery number. What is the mystery number? Could there be more than one? [Jan. 1997 (2)]

14.2. Multiply Ten. What is the smallest number that is a multiple of the first ten natural numbers? [Apr. 1997 (2)]

14.3. Divisors of 72. If two numbers divide 72, must their product also divide 72? Explain. [May 1995 (5)]

MAIN COURSES—*Robust problems to satisfy the multiples*

14.4. Factor a Million. How many pairs of whole numbers, neither of which has a zero in it, will have a product of 1 000 000? [Sept. 1994 (16)]

14.5. Rectangle Building. Bobby was using 1260 wooden square tiles, one inch on a side, to build rectangles. How many different perimeters can he get with rectangles that he can build that use all of the tiles? [Feb. 1995 (12)]

14.6. Magic Pairs. The numbers 3 and 37 037 are "magic." Here is how the magic works. Pick your favorite digit from among 1, 2, 3, … 8, 9 and multiply it by 3. Now multiply the result by 37 037. What is the final result? How many "magic" number pairs are there? [Sept. 1998 (10)]

14.7. Counting Zeros. How many terminal zeros are there at the end of the whole number 100!? (100 factorial is the product of all positive whole numbers less than or equal to 100.) [Oct. 1998 (7)]

DESSERTS—*Delectable morsels of enriched factors and multiples*

14.8. Tip the Driver. A group of friends took a bus trip. Each traveler gave the bus driver a tip using the same nine coins. The total tip was $8.41. How many dimes did the driver receive? [Apr. 1996 (16)]

SOLUTIONS TO PROBLEMS ABOUT FACTORS AND MULTIPLES I

*14.1. **Mystery Number.*** 60. Consider the prime factors of the numbers and look for the least common multiple. Three (3) and 5 are prime and $12 = (2)(2)(3)$. The LCM is $(2)(2)(3)(5) = 60$. Check to see that 60 has twelve factors. The factors are 1, 2, 3, 4, 5, 6, 10, 12, 15, 20, 30, and 60. Any multiple of 60 will have more than twelve factors, so 60 is the unique answer.

*14.2. **Multiply Ten.*** 2520. We need to have a multiple of 1, 2, 3, 4, 5, 6, 7, 8, 9, and 10 but not include a factor more than once. If a number is a multiple of 8, it will also be a multiple of 4 and 2. If it is a multiple of 9, it will also be a multiple of 3. So it must be a multiple of 8, 9, 5, and 7, but $8 \times 9 \times 5 \times 7 = 2520$.

*14.3. **Divisors of 72.*** Not always. A counterexample is 4 and 8. Each divide 72, but 32 does not.

*14.4. **Factor a Million.*** 64 and 15 625 is the only pair. $1\ 000\ 000 = 10^6 = 2^6 \times 5^6 = 64(15\ 625)$.

*14.5. **Rectangle Building.*** Eighteen different perimeters. List all factor pairs, using prime factorization of $1260 = (2^2)(3^2)(5)(7)$. The list is the following: (1, 1260), (2, 630), (3, 420), (4, 315), (5, 252), (6, 210), (7, 180), (9, 140), (10, 126), (12, 105), (14, 90), (15, 84), (18, 70), (20, 63), (21, 60), (28, 45), (30, 42), (35, 36). Corresponding to the pairs are the perimeters: 2552, 1264, 846, 638, 514, 432, 374, 298, 272, 234, 208, 198, 176, 166, 162, 146, 144, 142.

*14.6. **Magic Pairs.*** Sixteen magic pairs. A magic pair will be factors of 111 111. Since $111\ 111 = 3 \times 7 \times 11 \times 13 \times 37$, we can find the following pairs: (1, 111 111), (3, 37 037), (7, 15 873), (11, 10 101), (13, 8 547), (37, 3 003), (21, 5 291), (33, 3 367), (39, 2 849), (111, 1 001), (77, 1 443), (91, 1 221), (259, 429), (143, 777), (407, 273), (481, 231).

*14.7. **Counting Zeros.*** Twenty-four zeros. When the numbers from 100 to 1 are multiplied, zeros are created by combining fives and twos, since $5 \times 2 = 10$. Every multiple of 5 has at least one five, and four of them have two fives. Since there are twenty multiples of 5 and four multiples of 25, the number of zeros is $20 + 4 = 24$. There are plenty of factors of 2, since every other number has 2 as a factor.

*14.8. **Tip the Driver.*** Zero. The driver received $8.41 in equal amounts of nine coins from all riders. We search for factors of 841. We find $841 = 29 \times 29$, so we know that 29 riders must have each given tips of 29 cents. The only way to get 29 cents in nine coins is with five nickels and four pennies. Thus, no dimes were given.

15

Menu for Problems about Factors and Multiples II

APPETIZERS—*Light fare of factors and multiples*

15.1. Use All Digits. What is the largest multiple of 12 that can be written using each digit 0, 1, 2, 3, 4, 5, 6, 7, 8, and 9 exactly once? [May 1996 (5)]

15.2. Unusual Age. My age is a multiple of 7. Next year it will be a multiple of 5. When will this event occur again? [Nov. 1996 (3)]

MAIN COURSES—*Multiple foods to sustain thoughtful energy and effort*

15.3. Pricing Folders. A certain kind of folder was priced at 50 cents but found no buyers. The manager of the store decided to reduce the price. Within a few days she had sold her entire stock of these folders for $31.93. What was the reduced price of one folder? [Nov. 1994 (15)]

15.4. Find the Digits. Given that the number $5913d8$ is divisible by 12, what is the sum of all of the digits that could replace d? [Sept. 1997 (11)]

15.5. Area to Volume. The area of the floor of a rectangular room is 315 square feet. One wall is a rectangle whose area is 120 square feet, and another wall is a rectangle whose area is 168 square feet. If the ceiling and the floor are parallel, what is the volume of the room in cubic feet? [Sept. 1997 (14)]

15.6. Make a Product. You have ten number tiles containing the digits 0, 1, 2, 3, 4, 5, 6, 7, 8, and 9. Use seven tiles to make a true multiplication sentence, where each factor has two digits and the product has three digits. [Mar. 1995 (10)]

DESSERTS—*Delectable morsels to enrich the multiples*

15.7. Five Factors. List all numbers less than 100 with exactly five factors. Find the next two numbers, each greater than 100, that have exactly five factors. [May 1996 (16)]

15.8. All Digital Divisors. What is the least positive whole number that is divisible by all of the whole numbers from 1 through 9? [Dec. 1996 (7)]

SOLUTIONS TO PROBLEMS ABOUT FACTORS AND MULTIPLES II

15.1. *Use All Digits.* 9 876 543 120. Any number that uses all digits exactly once will be divisible by 3. We also need it to be divisible by 4. If the number named by the last two digits is divisible by 4, then so is the number. So we arrange the digits in order but reverse the positions of the 1 and 2 so that the last two digits name a number divisible by 4.

15.2. *Unusual Age.* In 35 years. List the multiples of 7 and the corresponding next years.

$7 \times$	7	14	21	28	35	42	49	56	63	70	77
$+1$	8	15	22	29	36	43	50	57	64	71	78

The multiples of 5 in the "next year" line are 15, 50, 85, …, and each of these is 35 years apart.

15.3. *Pricing Folders.* Either 31 cents or 1 cent; more likely 31 cents. The number 3193 is the product of the prime factors 31 and 103. So there were either 103 folders sold at 31 cents each or 3193 folders sold at 1 cent each.

15.4. *Find the Digits.* Four. We know that to be divisible by 12 requires divisibility by both 3 and 4. We first find the values of d that will make the number divisible by 4 and then eliminate any of those that will not make the number divisible by 3. To be divisible by 4, the number formed by the last two digits must be divisible by 4. So d can be 0, 2, 4, 6, or 8. To be divisible by 3, the sum of the digits must be divisible by 3. The sum of the digits is $26 + d$. So d could be 1, 4, or 7. The only value of d in both lists is 4.

15.5. *Area to Volume.* 2520 cubic feet. Represent the dimensions by L, W, and H. Since the greatest common factor (GCF) of 315 and 120 is 15, we know that W divides 15. Since the GCF of 315 and 168 is 21, we know that L divides 21. We also know that $WL = 315$, which means that $L = 21$ and $W = 15$. Then since $WH = 120$, we find $H = 8$. The volume is $21(15)(8) = 2520$.

15.6. *Make a Product.* Three possible solutions are as follows: $74 \times 13 = 962$, $62 \times 15 = 930$, and $52 \times 18 = 936$.

15.7. *Five Factors.* If a number has five factors, it must be of the form p^4 where p is a prime. So, $2^4 = 16$ and $3^4 = 81$ are the only numbers less than 100 that have five factors. The next numbers would be $5^4 = 625$ and $7^4 = 2401$.

15.8. *All Digital Divisors.* 2520. The number would have to be the lowest common multiple of the digits 1 through 9 or the product of $5 \times 7 \times 8 \times 9 = 2520$.

16

Menu for Find-a-Number Problems

APPETIZERS—*Light fare of hide-and-seek*

16.1. *Three Conditions.* Identify the number that satisfies all three of the following conditions:

(*a*) Is a composite between 62 and 72
(*b*) Sum of its digits is a prime number
(*c*) Has more than four factors. [Feb. 1996 (3)]

16.2. *A Decimal Number.* A decimal number rounds to 2. The product of its digits is 54, and the sum of its digits is 16. Find such a number. Can you find more than one number? [Jan. 1996 (4)]

16.3. *Two Digits.* What number is three less than a multiple of ten, two more than a perfect square, and has two digits? [Dec. 1996 (1)]

MAIN COURSES—*Mental food to sustain your searches*

16.4. *Jeff's Blocks.* Jeff had fewer than 1000 blocks. When he made five equal rows, he had one left over; with four equal rows, he had one left over; and with nine equal rows, he had none left over. How many blocks did Jeff have? [Jan. 1996 (14)]

16.5. *Ken's Shells.* Ken has between 3 and 100 shells in his collection. If he counts his shells three at a time, he has two left over. If he counts them four at a time, he has two left over. If Ken counts them five at a time, he still has two left over. How many shells does Ken have in his collection? [Apr. 1996 (6)]

16.6. *Becky's Shirts.* Becky put white, pink, and green T-shirts in the display case. One-third of the shirts were white. One-fourth of the shirts were pink. Ten shirts were green. How many shirts were in the case and how many were pink? [Sept. 1996 (11)]

DESSERTS—*Delectable morsels to enrich the spirit*

16.7. *Up and Down.* A number is increased by 50 percent, then the resulting number is decreased by 40 percent. What was the original number if the final number is eight less than the original? [Sept. 1996 (8)]

16.8. *Impossible Score.* What is the highest score below 100 that is impossible to score on a dart board where scores of 0, 4, or 7 are possible on a single throw? [Nov. 1997 (9)]

SOLUTIONS TO FIND-A-NUMBER PROBLEMS

16.1. Three Conditions. 70. The composites between 62 and 72 are 63, 64, 65, 66, 68, 69, 70. Those having a prime for sum of digits are 65, 70. Sixty-five (65) has 1, 5, 13, and 65 as factors. Seventy (70) has 1, 2, 5, 7, 10, 14, 35, and 70 as factors.

16.2. A Decimal Number. One answer is 1.96, since $9 \times 6 \times 1 = 54$, $9 + 6 + 1 = 16$, and 1.96 rounds to 2. Some other possibilities are 2.1319, 2.3911, 1.69, 1.9321. Can you find others?

16.3. Two Digits. 27. All two-digit numbers that are three less than a multiple of ten are 17, 27, 37, 47, 57, 67, 77, 87, and 97. Numbers that are two less are 15, 25, 35, 45, 55, 65, 75, 85, and 95. Only 25 is a square.

16.4. Jeff's Blocks. Jeff could have 981 or 801 or 621 or 441 or 261 or 81 blocks. The numbers that have remainders of 1 when divided by 4 or 5 have a units digit of 1 and an even tens digit. Numbers divisible by 9 have a sum of digits that is divisible by 9. Six numbers less than 1000 meet those conditions.

16.5. Ken's Shells. Sixty-two shells. Any number divisible by 5 ends in a 0 or 5, so a number leaving a remainder of 2 after division by 5 must end in a 2 or 7. The numbers less than 100 that satisfy this condition are 7, 12, 17, 22, 27, 32, 37, 42, 47, 52, 57, 62, 67, 72, 77, 82, 87, 92, and 97. Those which leave a reminder of 2 when divided by 3 are 17, 32, 47, 62, 77, and 92. Finally, check to see which of these leave a remainder of 2 when divided by 4. A more direct approach is to note that two more than the number must be divisible by 3, 4, and 5; so two more than the number must be divisible by 60. Sixty-two (62) is the only number less than 100 that meets those conditions.

16.6. Becky's Shirts. Twenty-four shirts were in the case, and six were pink. Since 1/3 were white and 1/4 were pink, the remaining shirts must be green. Since $1/3 + 1/4 = 7/12$, then 5/12 must be green. Knowing ten shirts were green leads to $10(12/5) = 24$ shirts in the case. Since $(1/4)(24) = 6$, there must be six pink shirts.

16.7. Up and Down. 80. Start with 1; increase by 50 percent to get 3/2. Then decrease by 40 percent to get 6/10 of $3/2 = 9/10$. The final number is 9/10 of the original number. But 1/10 of the original number is 8, so the original number is 80. This problem can also be solved by algebra and by guess-check-revise.

16.8. Impossible Score. 17. It is possible to make every number greater than 17 with combinations of 4s and 7s. If you have a 7, you can trade for two 4s to get an increase of 1. If you have five 4s, you can trade for three 7s to get an increase of 1. If you have over 17, you will always have either a 7 to trade or five 4s to trade.

17

Menu for Problems on Geometry

APPETIZERS—*Light fare of geometric morsels*

17.1. Rectangles. How many rectangles are in the array pictured at right? [May 1994 (5)]

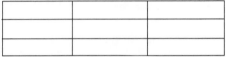

Figure for problem 17.1

17.2. Three Straws. Two straws are taped end to end so that they can be bent. They may form either 90-degree or 180-degree angles. No other angles are permitted. Only two formations are possible (see margin). How many formations are possible when three straws are taped end to end? [Jan. 1995 (3)]

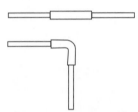

Figure for problem 17.2

17.3. Find Triangles. How many triangles are in the regular pentagon drawn with all of its diagonals? (See figure in the margin.) [Feb. 1998 (3)]

MAIN COURSES—*More hearty fare*

17.4. Big Wheel, Little Wheel. How many more complete turns will a wheel with a circumference of 3 feet make than a wheel with a circumference of 4 feet in traveling one mile? [Apr. 1994 (13)]

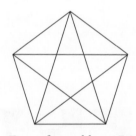

Figure for problem 17.3

17.5. Change in Area. The length of a rectangle is increased by 20 percent, and its width is decreased by 10 percent. By what percent does the area of the rectangle change? What if, instead, we decreased the length by 20 percent and increased the width by 10 percent? [Oct. 1996 (15)]

17.6. Color the Logo. Consider the NCTM logo in the margin. Color each triangular and circular part of the logo so that no adjacent pieces have the same color. Can you color it with two colors? Three colors? Four colors? Explain. [Apr. 1998 (13)]

DESSERTS—*Pleasant repast*

17.7. Shady Triangles. A square region is partitioned into eight congruent triangular regions by its four lines of symmetry. In how many ways can you shade four of the triangular regions? Do not count congruent patterns as different. [Feb. 1995 (18)]

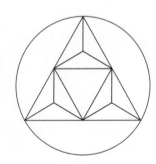

Figure for problem 17.6

17.8. Patio Bricks. Patio bricks are three inches by six inches. Plastic frames that are nine inches by twelve inches are available that hold six such bricks. Sketch as many ways of laying six whole bricks in the frame as you can. [Apr. 1996 (19)]

17.9. Hidden Rectangles. How many distinct rectangles of any size are in the figure in the margin? The figure is of six congruent rectangles packed end to end. [Feb. 1998 (10)]

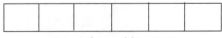

Figure for problem 17.9

SOLUTIONS TO PROBLEMS IN GEOMETRY

17.1. Rectangles. Thirty-six rectangles. There are nine small rectangles; twelve "1 × 2" or "2 × 1"; four "2 × 2"; six "1 × 3" or "3 × 1"; four "2 × 3" or "3 × 2"; and one "3 × 3." 9 + 12 + 4 + 6 + 4 + 1 = 36.

17.2. Three Straws. Four formations are possible. (See margin.)

17.3. Find Triangles. Thirty-five (35) triangles. Seven unique—that is, not congruent to any other triangle in the figure—triangles are in the figure. The five-sided rotational symmetry results in five congruent copies of each.

17.4. Big Wheel, Little Wheel. 440 more turns. The smaller wheel will make 5280/3 = 1760 revolutions. The larger wheel will make 5280/4 = 1320 revolutions. The difference is 440 revolutions.

17.5. Change in Area. Eight percent increase in the first case and 12 percent decrease in the second case. Let L represent the length and W represent the width, so LW represents the area. If L is increased by 20 percent, it becomes $1.2L$. If W is decreased by 10 percent, it becomes $0.9W$. The area of the new rectangle is $(1.2L)(0.9W) = 1.08LW$, which is an 8 percent increase over the original LW. In the second case, we have a new area of $(0.8L)(1.1W) = 0.88LW$, which accounts for a 12 percent decrease from LW.

17.6. Color the Logo. The logo cannot be colored with only two colors, since three small triangles meet in a single point. So, a minimum of three colors is needed. In fact, the logo can be colored with three colors (and thus also with four colors).

17.7. Shady Triangles. Thirteen ways.

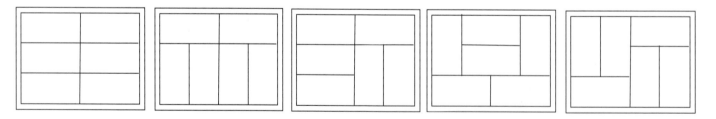

17.8. Patio Bricks. Five ways are pictured.

17.9. Hidden Rectangles. Twenty-one (21) rectangles. Six single rectangles, five composed of two small rectangles, four composed of three small rectangles, three composed of four small rectangles, two composed of five small rectangles, and one composed of six small rectangles. 6 + 5 + 4 + 3 + 2 + 1 = 21.

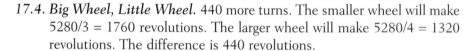

Figure for solution 17.2

18
Menu for Max–Min Problems

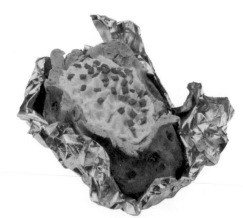

APPETIZERS—*Light fare to optimize your appetite*

18.1. *Miniproduct.* A miniproduct is the smallest possible product when multiplying a three-digit number by a two-digit number. Find the miniproduct using each of the digits 1, 2, 3, 4, and 5 to make up your two factors. A superproduct is the largest possible product when multiplying a three-digit number by a two-digit number. Find the superproduct using each of the digits 1, 2, 3, 4, and 5 to make up your two factors. [May 1994 (9)]

18.2. *Close but Less.* The sentence 6.5 × 4.321 = 28.0865 has each of the digits 1, 2, 3, 4, 5, and 6 exactly once on the left-hand side. The right-hand side is close to, but less than, 30. Can you find another multiplication sentence that contains each digit 1, 2, 3, 4, 5, and 6 exactly once on the left side and is closer to 30 than 28.0865? What is the closest you can get to 30 and stay under 30? [Nov. 1997 (4)]

18.3. *Just Close.* Change the rules above so that you can have a product greater than 30 but still get as close to 30 as possible. Use each digit 1, 2, 3, 4, 5, and 6 exactly once. Can you get closer to 30 using these new rules? [Nov. 1997 (5)]

MAIN COURSES—*Mental food to sustain maximal energy*

18.4. *Progressive Taxes.* The city of Taxaphobia imposes a progressive tax rate; the more you earn, the higher your tax rate. For incomes that are less than $100 000, the percent tax rate is the number of thousands of dollars that a taxpayer earns. For example, those who earn $13 000 are taxed at 13 percent. Those earning $60 000 pay 60 percent. Under this system, what salary, to the nearest $1 000, would you need to earn so that your take-home pay would be as high as possible? [Nov. 1997 (11)]

18.5. *Wait in Line.* Your group of fifty is in a single line waiting to ride a roller coaster at Great America Amusement Park. The coaster has eleven cars, which hold two people each. The ride lasts four minutes, and three minutes are required to unload one group and load the next group. What is the least amount of time from when the first member of your group begins to load until the last member of your group unloads? What is the greatest amount of time? [May 1998 (11)]

18.6. *Two Sales Plans.* A company offers two options for sales partners: $150 plus a 6 percent commission weekly or a flat 11 percent commission. Mrs. Lee, the top salesperson in the company, averages $3500 a week in sales. Which option should she take? Why? [May 1998 (14)]

DESSERTS—*Delectable morsels for maximum enrichment*

18.7. *Votes for B.* Five candidates—A, B, C, D, and E—were in an election. Candidate A was elected with 30 votes, B placed second, and E was last with three votes. If 49 votes were cast and no two candidates had the same number of votes, what is the smallest amount of votes that B could have received? [May 1996 (18)]

18.8. *Maximize a Product.* The values of *a*, *b*, *c*, and *d* are 1, 2, 3, and 4—but not necessarily in that order. Find the largest possible value of $ab + bc + cd + da$. [Oct. 1996 (6)]

SOLUTIONS TO MAX–MIN PROBLEMS

18.1. Miniproduct. Mini = 245 × 13 = 3 185; super = 431 × 52 = 22 412. These answers can be found by trying all reasonable cases. You also may observe $(abc)(de) = 1\,000ad + 100(ae + bd) + 10(be + dc) + ec$. For mini, you want the ad to be minimal; for super, you want the ad to be maximal.

18.2. Close but Less. Either 65.1 × 0.432 = 28.1232 or 43.2 × 0.651 = 28.1232. A guess-and-check strategy works well.

18.3. Just Close. Yes, you can get closer. 6.1234 × 5 = 30.617.

18.4. Progressive Taxes. $50 000. You can construct a table with money earned in multiples of $10 000, the taxes assessed, and the take-home pay. You can see symmetry in that the take-home pay for $X is the same as the take-home pay for $100 000 – X.

18.5. Wait in Line. Twenty-four minutes minimum and thirty-one minutes maximum. Twenty-two people can ride at one time. Then, three minutes are required to load the first set of people. After the initial loading, each trip, which includes the ride and the unloading process, takes seven minutes. Two full trips will be needed to carry forty-four people. The minimum time will occur when the remaining six go together. (3 + 3 × 7 = 24). The maximum time will occur when the remaining six are in two separate trips. (3 + 4 × 7 = 31). For example, if the first person in the group is the last one to board, then a fourth trip will be required for the last five people in line.

18.6. Two Sales Plans. The choice depends on the amount of sales. If sales are under $3000, then the first option is better. If sales are greater than $3000, then the second choice is preferred.

18.7. Votes for B. Seven votes. Let b, c, and d represent the number of votes that B, C, and D received. Since 49 votes were cast, $b + c + d = 49 - 33 = 16$ votes. Since candidate D got more than three votes, try $d = 4$. Then C had to get at least five votes, which would leave seven votes for B. If B were to get less than seven votes, say six votes, then the maximum number of votes cast would have been 30 + 6 + 5 + 4 + 3 = 48.

18.8. Maximize a Product. 25. Note that $ab + bc + cd + da = b(a + c) + d(c + a) = (a + c)(b + d)$. The possible products are 3 × 7, 4 × 6, and 5 × 5. The last of these is maximal.

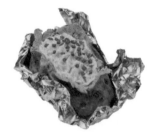

19

Menu for Problems about Measurement I

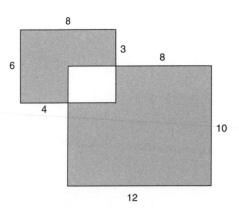

Figure for problem 19.2

APPETIZERS—*Light fare to measure your appetite*

19.1. *Around a Wheel*. A wheel with radius of one inch rolls (without slipping) around another wheel with a radius of six inches and returns to its original position. How many rotations will the smaller wheel make? [Dec. 1994 (2)]

19.2. *Area*. Determine the area of the shaded region shown at right. (Try to determine the area in as many different ways as you can.) [Jan. 1995 (5)]

19.3. *Angle Size*. In the figure in the margin, equilateral triangle *TOP* rests on top of square *TOWN*. What is the measure of angle *POW*? What is the measure of angle *PON*? Of angle *PWO*? [Mar. 1997 (1)]

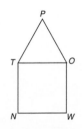

Figure for problem 19.3

MAIN COURSES—*Robust problems to satisfy the hungry measurer*

19.4. *Area to Length*. The areas of three faces of a rectangular solid are 54, 72, and 108 square units. Find the whole number measures of the edges. [Apr. 1994 (11)]

19.5. *Prisms in a Box*. How many 3″ × 3″ × 1″ rectangular prisms fit into a 4″ × 4″ × 4″ box? [Feb. 1998 (14)]

DESSERTS—*Delectable morsels from the world of measurement*

19.6. *Compare Cylinders*. Take two sheets of notebook paper that are the same size. Roll one sheet vertically and the other horizontally and tape each, without overlapping, to form a right circular cylinder. If you filled each with popcorn, would they contain the same amount? Explain. [Apr. 1995 (18)]

19.7. *Area of Pieces*. Each figure (see margin) is a rectangle with an area of 1. What is the area of each numbered region? [In (a), the vertices divide the sides into thirds. In (b), one vertex of the triangle is at the midpoint of the rectangle. In (c), one vertex of triangle 1 is at the midpoint of the rectangle. In (d), the vertices of the triangle are at midpoints of edges of the rectangle.] [Dec. 1995 (17)]

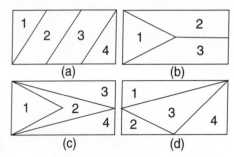

Figure for problem 19.7

19.8. *Area Ratio*. Quadrilateral *ABCD* in the margin is a rectangle with length to width in the ratio 3:2. Segments *AN* and *BM* are bisectors of angles *A* and *B*, respectively. What is the ratio of the area of the shaded region to the area of *ABCD*? [Feb. 1998 (7)]

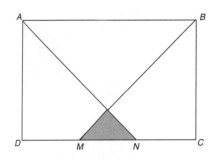

Figure for problem 19.8

SOLUTIONS TO PROBLEMS ABOUT MEASUREMENT I

19.1. *Around a Wheel.* Six rotations. The circumference of the small wheel is 2π. The circumference of the large wheel is 12π. Since $12\pi/2\pi = 6$, it will take six rotations.

19.2. *Area.* 144. $(6 \times 8) + (10 \times 12) - 2(4 \times 3)$. Or push down the top shaded hexagon three units and then add $(12 \times 10) + (4 \times 6)$. Or push the top hexagon to the right and then add $(10 \times 12) + (3 \times 8)$. Several other strategies are possible.

19.3. *Angle Size.* 150 degrees, 105 degrees, and 15 degrees. In a square, all the angles measure 90 degrees; in an equilateral triangle, all angles measure 60 degrees. $POW = POT + TOW = 60 + 90 = 150$. $PON = POT + TON = POT + 0.5(TOW) = 60 + 45 = 105$. POW is isosceles, so $PWO = 0.5(180 - POW) = 0.5(180 - 150) = 0.5(30) = 15$.

19.4. *Area to Length.* The edges measure 6, 9, and 12. Let the dimensions be L, W, H and $LW = 108$, $LH = 54$, and $WH = 72$. Then, $LW = 108 = 2(54) = 2LH$. So, $W = 2H$. Substituting, we get $2HH = 72$ and then $H = 6$. Substituting $H = 6$ into $LH = 54$, we get $L = 9$. Substituting $L = 9$ into $LW = 108$, we get $W = 12$. This can also be done by trial and error, looking for common factors.

19.5. *Prisms in a Box.* Six prisms. A $3'' \times 3'' \times 3''$ prism has volume of 9 cubic inches. A $4'' \times 4'' \times 4''$ prism has volume of 64 cubic inches. So no more than seven of the smaller prisms can fit inside the larger prism. If you stack four of the smaller prisms, you can fill a $3'' \times 3'' \times 4''$ space. There remains room for only two more pieces to be stacked in the larger prism.

19.6. *Compare Cylinders.* No. One cylinder will have a base circumference of 11 inches (a base radius of $11/2\pi$, about 1.75 inches) and a height of 8.5 inches. Its volume will be $\pi(11/2\pi)^2(8.5)$, which is about 82 cubic inches. The other cylinder will have a base circumference of 8.5 inches (a base radius of $8.5/2\pi$, about 1.35 inches) and a height of 11 inches. Its volume will be $\pi(8.5/2\pi)^2 11$, which is about 63 cubic inches.

19.7. *Area of Pieces.* In (a), each triangle has an area of 1/6, and the parallelograms each have an area of 1/3. In (b), the triangle has an area of 1/4 and the two trapezoids each have areas of 3/8. In (c), each of the four regions has an area of 1/4. In (d), the lower left triangle has an area of 1/8; the upper left triangle has an area of 1/4; the lower right triangle has an area of 1/4; the remaining triangle has an area of 3/8.

19.8. *Area Ratio.* The ratio is 1:24. Let $AB = 3$ and $AD = 2$. Then isosceles right triangles ADN and BCM have an area of $(1/2)(2)(2) = 2$ square units. Label the intersection of AN and BM as E. Isosceles right triangle AEB has an area of $(1/2)(3)(3/2) = 9/4$ square units. The area of $ABCD$ is $(3)(2) = 6$ square units. You can find the area of MEN because area ADN + area BCM + area AEB − area MEN = area $ABCD$. Alternatively, divide the rectangle into six congruent squares. The area to be determined is 1/4 of the area of one of those six congruent squares.

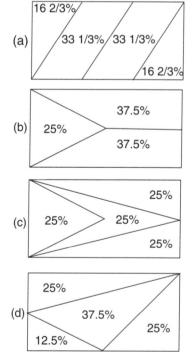

(a) 16 2/3% 33 1/3% 33 1/3% 16 2/3%

(b) 25% 37.5% 37.5%

(c) 25% 25% 25% 25%

(d) 25% 37.5% 12.5% 25%

Figure for solution 19.7

20

Menu for Problems about Measurement II

APPETIZERS—*Light fare to measure your skill*

20.1. Shortcut. If a field is 14 meters × 48 meters, how many meters will you save by running diagonally across the field instead of running along two adjacent sides? [Apr. 1995 (4)]

20.2. Tree Replacement. In New York City, people who damage a tree have to follow one of two replacement rules. Rule 1: The total diameter of the new trees planted must equal that of the tree that was damaged. Rule 2: The total area of the cross sections of the new trees must equal the area of the cross section of the damaged tree. Explain which rule, if either, is more fair. [Dec. 1995 (1)]

MAIN COURSES—*Mental food to sustain a measured effort*

20.3. Partitioned Rectangle. A rectangle (see figure below) is divided into four smaller rectangles with the upper-right rectangle being a square. Find the area of the shaded rectangle if the three remaining rectangles have areas of 95, 25, and 228 square feet, respectively. [Apr. 1996 (11)]

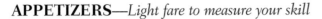

20.4. Freeway Poles. Freeway poles are placed at regular intervals. The distance from the first pole to the fifth pole is 300 feet. What is the distance from the tenth pole to the twentieth pole? [Nov. 1996 (16)]

20.5. Unit Triangle. Use the smallest triangle in the NCTM logo on the right as a unit. Find the area of as many other polygons embedded in the figure as you can using the smallest triangle as a unit. [Mar. 1998 (9)]

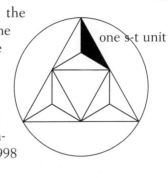

one s-t unit

DESSERTS—*Delectable morsels to measurably enrich your spirit*

20.6. Pizza Comparison. A waiter in a pizza restaurant said that its medium pizza was 10 inches in diameter and its large pizza was 15 inches in diameter. He also said that the 15-inch pizza was about twice as large as the 10-inch pizza. Was he correct? [Dec. 1994 (16)]

20.7. Circular Region. Points *A* and *B* below are the centers of two congruent circles that are tangent externally and packed in a rectangle. If *AB* = 10, find the total area of the shaded region. Approximate π as 3.14 and approximate the area to the nearest enth of a square unit. [Apr. 1997 (8)]

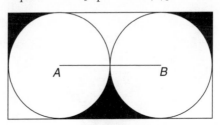

20.8. Unpainted Block. A solid block of wood in the shape of a right rectangular prism and measuring 4″ × 5″ × 8″ is dipped into red paint. It is then cut into one-inch cubes. How many of the one-inch cubes will have no paint on them? [Jan. 1998 (9)]

SOLUTIONS TO PROBLEMS ABOUT MEASUREMENT II

20.1. *Shortcut.* About <u>12 meters.</u> Use the Pythagorean theorem to measure the diagonal as $\sqrt{14^2 + 48^2}$, which is a little less than 50. Running around two sides would be 62 meters, a difference of about 12 meters.

20.2. *Tree Replacement.* Rule 2 comes closer to replacing the volume of the damaged tree. Use an example. Suppose you damage a tree having a diameter of 5 inches and replace it with five trees with diameters of 1 inch each. The cross-sectional area damaged is $\pi(2.5)^2$, or about 19.6 square inches. The cross-sectional area replaced is $(5)\pi(0.5)^2$, or about 3.9 square inches.

20.3. *Partitioned Rectangle.* Sixty square feet. Since the upper-right corner is a square with an area of 25 square feet, each side is 5 feet in length. The rectangle with an area of 95 square feet must then be 5 ft × 19 ft. The rectangle with an area of 228 square feet must then be 19 ft × 12 ft. The remaining rectangle must be 12 ft × 5 ft and would have an area of 60 square feet.

20.4. *Freeway Poles.* 750 feet. There are four intervals between the first and fifth pole, so each interval is 300/4 = 75 feet. There are ten intervals between the tenth and twentieth pole, so that 10(75) = 750 feet.

20.5. *Unit Triangle.* The large triangle has an area of 12; the medium triangle has an area of 3; the central hexagon has an area of 6.

20.6. *Pizza Comparison.* The waiter was correct. The 15-inch pizza has an area of $\pi(7.5)^2$ square inches, and the 10-inch pizza has an area of $\pi(5)^2$ square inches. The ratio of the areas is 56.25/25, which is greater than 2/1.

20.7. *Circular Region.* About 21.5 square units. Since $AB = 10$, the radius of each circle would be 5. Therefore the length of the rectangle is 20, and the width is 10. So the area of the rectangle is 200, and the area of the circles is $2\pi r^2$ where $r = 5$. The approximate area of the circles is then $2(3.14)(25) = 157$. The approximate shaded area is then $(0.5)(200 - 157) = 21.5$.

20.8. *Unpainted Block.* Thirty-six cubes. The original block contains $4 \times 5 \times 8 = 160$ blocks. After the block is dipped in paint, only the inside blocks will have no paint. A one-inch layer would have to be cut all around to see the unpainted cubes. The unpainted block would contain $(4 - 2)(5 - 2)(8 - 2) = 2 \times 3 \times 6 = 36$ cubes.

21
Menu for Problems about Patterns

APPETIZER—*Light fare to whet your appetite for patterns*

21.1. Nine Stones. Nine stones are arranged in a straight line. They are counted from left to right as 1, 2, 3, …, 9 and then from right to left so that the stone previously counted as 8 is counted as 10. The pattern is continued to the left until the stone previously counted as 2 is counted as 18, 3 as 19, and so on. The counting continues in this way. Which of the original stones is counted as 99? Express your answer as the first number assigned to that stone. [May 1997 (2)]

21.2. Triangular Array. What numeral will be listed directly beneath 25 when this triangular array is continued? [Sept. 1997 (2)]

```
 1
 2    3
 4    5    6
 7    8    9   10
11   12   13   14   15
```

21.3. Hundreds Chart. Here is part of a hundred chart.

1	2	3	4	5	6	7	8	9	10
11	12	13	14	15	16	17	18	19	20
21	22	23	24	25	26	27	28	29	30
31						37	38		

Without completing the chart, determine which number should replace *x* in this diagram? [Sept. 1998 (4)]

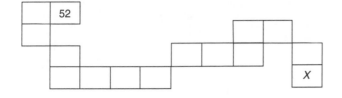

MAIN COURSES—*Robust patterns to satisfy the insightful*

21.4. Powers. Find the units digit of n^{100} for each of the values $n = 2, 3, 4, 5, 6$, and 7. Look for a pattern involving smaller powers of each base. [May 1995 (14)]

21.5. Guess Next. Find a pattern and predict what comes next in the sequence that begins 77, 49, 36, 18, … [Sept. 1996 (10)]

21.6. Odd Triangle. Consider the following pattern:

```
              1
          3       5
       7     9      11
    13   15    17    19
  21   23   25   27   29
```

Find a pattern and state a rule that relates the difference between the last number in a row and the first number in a row with the row number. Find a pattern and state a rule that gives the sum of all numbers in a row in terms of the row number. [Feb. 1997 (14)]

DESSERTS—*Delectable patterns of morsels*

21.7. Toothpick Squares. Toothpicks are used to build a rectangular grid that is twenty toothpicks long, ten toothpicks wide, and filled with squares one toothpick on a side. What is the total number of toothpicks used? If *a* represents the number of toothpicks in the length and *b* represents the number of toothpicks in the width, write an expression representing the total number of toothpicks in the figure. [May 1996 (19)]

21.8. Shading Squares. All the outside squares of a 3 × 3 square array are shaded. One square is not shaded. If all the outside squares of an *m* × *m* square array were shaded, express the number of shaded squares in terms of *m*. [Feb. 1997 (7)]

SOLUTIONS TO PROBLEMS ABOUT PATTERNS

21.1. Nine Stones. 3. Begin a list and look for a pattern.

1	2	3	4	5	6	7	8	9
17	16	15	14	13	12	11	10	
	18	19	20	21	22	23	24	25
33	32	31	30	29	28	27	26	…

Note that alternating rows differ by 16 (consider columns 2 through 8 only). Since $99 = 16(6) + 3$, the number 99 should be in the sixth alternate row under 3.

21.2. Triangular Array. 32. You can continue the array or look for a pattern. The difference between numbers in the first row and the second row is 1; the difference between the numbers in the second and third rows is 2. The difference between the numbers in the nth row and the $(n + 1)$st row is n. The number 25 is in the 7th row, so the number beneath it is $25 + 7 = 32$.

21.3. Hundreds Chart. $x = 90$. The left-hand corner is 51. The x is in the row containing 81 through 90. It is nine positions to the right of the 51, so it is $81 + 9 = 90$.

21.4. Powers.

For n	= 2	3	4	5	6	7
units digit of n^{100}	= 6	1	6	5	6	1

For example, successive powers of 2 have units digits 2, 4, 8, 6, and 2, in cycle. Every fourth member of the cycle will be 6; and the 100th member of the pattern will be 6.

21.5. Guess Next. 8. Each term appears to be the product of the digits of the previous term.

21.6. Odd Triangle. The difference is $2n - 2$. The sum is n^3. Making a table will help to show the pattern.

21.7. Toothpick Squares. 430. There will be $20 \times 11 = 220$ toothpicks in rows and $10 \times 21 = 210$ toothpicks in columns, for a total of $220 + 210 = 430$ toothpicks. In general, there will be $a(b + 1)$ toothpicks in rows and $b(a + 1)$ toothpicks in columns, for a total of $a(b + 1) + b(a + 1) = 2ab + a + b$ toothpicks.

21.8. Shading Squares. $m^2 - (m - 2)^2$ or $4m - 4$. You can make a table and look for a pattern. Or consider that there are m shaded squares on each of the four edges of the figure. Since the four corner squares are counted twice, they must be subtracted from the total, leaving $4m - 4$.

22
Menu for Problems about Probability I

APPETIZERS—*Probably light fare*

22.1. *Five Spades.* From a standard deck of fifty-two cards, how many cards would you have to draw, without looking at them, to be absolutely certain (a probability of 1) that you had five spades? [Jan. 1995 (2)]

22.2. *Boys and Girls.* In a family of four children, which is more common, an equal or an unequal number of boys and girls (or are both equally common)? Assume the probability of a boy birth is 1/2. [Dec. 1998 (4)]

22.3. *Shake a Three.* What is the probability of rolling a multiple of 3 with a pair of fair dice? [May 1996 (2)]

MAIN COURSES—*Mental sustenance*

22.4 *Bags and Balls.* Bag 1 contains three red balls and two green balls. Bag 2 contains two red balls and one green ball. To win a prize, you must pick a red ball without looking in either bag. Suppose you decide to flip a fair coin; if the coin is "heads," you draw a marble from Bag 1; if "tails," you draw a marble from Bag 2. What is your probability of winning? [Nov. 1994 (10)]

22.5. *Sum and Product.* Two different numbers are selected from the set of positive integers less than or equal to 5. What is the probability that the sum of the two numbers is greater than their product? Express your answer as a common fraction. [May 1997 (15)]

DESSERTS—*Probable reward for mental effort*

22.6. *Weighted Cube.* The faces of a weighted cube are numbered 1, 2, 3, 4, 5, and 6, with one digit on each face. When tossed, the cube is twice as likely to land with an even digit showing on its top face as an odd. Each odd number is as likely to appear as any other odd number, and each even number is as likely to appear as any other even number. What is the probability that the cube, when tossed, will show a 5 on its top face? [Sept. 1997 (8)]

SOLUTIONS TO PROBLEMS ABOUT PROBABILITY I

22.1. *Five Spades.* Forty-four cards. You could draw all of the hearts, diamonds, and clubs (3×13) before drawing the fifth spade. You may need to draw $39 + 5 = 44$ cards to ensure five spades.

22.2. *Boys and Girls.* Unequal is more common. There are sixteen strings of B and G. The six strings—BBGG, BGBG, BGGB, GBBG, GBGB, and GGBB—are those having equal numbers of Bs and Gs. Probability of unequal is 10/16.

22.3. *Shake a Three.* 1/3. There are 36 different equally likely pairs that can be rolled. Twelve pairs add to a multiple of three: (1, 2), (2, 1), (1, 5), (5, 1), (2, 4), (4, 2), (3, 3), (4, 5), (5, 4), (3, 6), (6, 3), and (6, 6). The probability is 12/36 = 1/3.

22.4. *Bags and Balls.* 19/30. The outcomes can be identified as (Bag 1, Red), (Bag 1, Green), (Bag 2, Red), (Bag 2, Green). The probabilities are $(1/2)(3/5) = 3/10$, $(1/2)(2/5) = 2/10$, $(1/2)(2/3) = 1/3$, and $(1/2)(1/3) = 1/6$. The first and third of these result in a win, so the probability of a win is $3/10 + 1/3 = 19/30$.

22.5. *Sum and Product.* 8/20 or 2/5. Make a chart of the twenty possible pairs (assuming the numbers are chosen without replacement, so two distinct numbers are chosen). The sum will be greater than the product for the four pairs—(1, 2), (1, 3), (1, 4), (1, 5)—and the four pairs formed by reversing the order of these four.

22.6. *Weighted Cube.* 1/9. Let the probability of getting a 1 be x. Then the probability of getting a 3 is x; the probability of getting a 5 is also x. The probability of getting a 2 is $2x$; of getting a 4 is $2x$; of getting a 6 is $2x$. The sum of all probabilities associated with the six faces is 1. So, $x + 2x + x + 2x + x + 2x = 1$ and so $9x = 1$, from which $x = 1/9$.

23
Menu for Problems about Probability II

APPETIZERS—*Problems to whet your probability appetites*

23.1. Scratch Off. Bob's Burger Barn is having a contest. Each customer is given a card with five gold spots. Each spot will show the name of a food item when its coating is scratched off. Each card has only two matching spots. Customers can scratch off only two spots. Bob thinks this is producing too many winners. If Bob wants each player's probability of winning to be about 1/25, how many spots should he add to the cards? [Nov. 1995 (4)]

23.2. Red, Green, and Blue. A bag contains five red marbles, three green marbles, and four blue marbles. How many green marbles must be added so that the probability of drawing a green marble is 3/4? [Apr. 1997 (1)]

23.3. Shaded Triangles. If a triangle is randomly selected from the triangles pictured in the margin, what is the probability that at least part of the interior of that triangle is shaded? [May 1997 (4)]

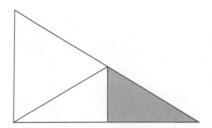

Figure for problem 23.3

MAIN COURSES—*Robust problems to satisfy the hearty appetite*

23.4. Toss a Tetrahedron. A regular tetrahedron has four faces that are congruent equilateral triangles. One face is colored blue, one red, one green, and one yellow. If you toss the tetrahedron twice, what is the probability that the figure will land with the same color face down both times? [Oct. 1994 (13)]

23.5. Multiply a Pair of Dice. Toss a pair of dice and multiply the number of dots showing on the uppermost faces. What is the probability of getting a product that is a multiple of 3? A multiple of 5? A multiple of 3 or 5? A multiple of 3 and 5? [Nov. 1994 (13)]

23.6. Center or Edge. If an arrow has an equal chance of landing at any point on a circular target, is it more likely to land closer to the center or to the edge? Explain your reasoning. [Nov. 1997 (13)]

DESSERTS—*Delicacies from probability*

23.7. Terminators. If the numerator of a fraction is randomly selected from the set {1, 3, 5, 7, 9} and the denominator is randomly selected from the set {1, 2, 3, 4, 5}, what is the probability that the decimal representation of the fraction is a terminating decimal? [Sept. 1997 (9)]

SOLUTIONS TO PROBLEMS ABOUT PROBABILITY II

23.1. *Scratch Off.* Add two or three spots. Currently, there is a 1 in (5)(4)/2 = 10 chance of winning. With seven spots, there is 1 in (7)(6)/2 = 21. With eight spots, there is 1 in (8)(7)/2 = 28. If Bob adds two spots, then there is greater than 1/25 chance of winning; if he adds three spots, there is a little less than 1/25 chance of winning.

23.2. *Red, Green, and Blue.* Twenty-four green marbles. If green marbles represent 3/4, then the nine red and blue marbles represent 1/4. If 9 represents 1/4, then 27 represents 3/4. To get to 27 from 3, you must add 24.

23.3. *Shaded Triangles.* 3/5. There are five triangles: two are partly shaded, one is entirely shaded, and two are not shaded.

23.4. *Toss a Tetrahedron.* 1/4. There are sixteen equally likely color pairs. Of these, BB, RR, GG, and YY represent the same color occurring. So the probability is 1/4. You can also reason that regardless of what is thrown the first time, there is a 1/4 probability of getting that same color again.

23.5. *Multiply a Pair of Dice.* 5/9, 11/36, 3/4, 1/9. Make a table of the 36 number pairs and find the product in each case or reason using knowledge of factors. To get a multiple of 3, you need at least one die to land a 3 or a 6. There are ten pairs that contain exactly one 3 and eight pairs that contain exactly one 6, but no 3, and one pair (3, 3) and one pair (6, 6). So 20/36 of the products will be a multiple of 3. To get a multiple of 5, you need at least one die to land a 5. There are ten pairs with exactly one five and the pair (5, 5), making eleven pairs having products that are multiples of 5. To be a multiple of 3 and 5, you must have a 3 or a 6 as one member of the pair and a 5 as the other member of the pair. There are four such pairs, so the probability is 4/36 = 1/9. To get the probability of being a multiple of 3 or 5, you can note that there are twenty pairs that are multiples of 3 and eleven pairs that are multiples of 5, with four pairs in both lists. So the probability of getting a multiple of 3 or 5 is 20/36 + 11/36 − 4/36 = 27/36 = 3/4.

23.6. *Center or Edge.* Closer to the edge. Suppose the target has a radius of 4 inches. Draw a concentric circle with a radius of 2 inches. The probability of landing closer to the center is the area of the smaller circle compared with the area of the larger circle. This is $4\pi/16\pi = 1/4$. So you are three times as likely to land closer to the edge.

23.7. *Terminators.* 22/25. There are five possible numerators and five denominators, so twenty-five fractions can be formed. Lowest-terms fractions with a denominator of 1, 2, 4, or 5 will be equivalent to terminating decimals. Thus, we need only check the five fractions with a denominator of three. Of these, 1/3, 5/3, and 7/3 do not terminate; 3/3 and 9/3 reduce to 1 and 3, respectively, and consequently do terminate.

24

Menu for Problems about Proportions

APPETIZERS—*Light fare of the proper proportion*

24.1. *Jose's Shadow.* Jose's shadow is 8 feet long at the same time that the shadow of a nearby tree is 32 feet long. If Jose is 6 feet tall, how tall is the tree? [Apr. 1994 (6)]

24.2. *Zips and Zaps.* One zip weighs as much as three zaps. Two zaps weigh as much as five zowwies. Three zowwies weigh as much as two swooshes. If one swoosh weighs sixty pounds, how many pounds does one zip weigh? [Jan. 1998 (2)]

24.3. *Cathy's Cheesecake.* Cathy's cheesecake recipe calls for 3 packages of cream cheese, 3/4 cup of sugar, 3 eggs, 1 teaspoon of vanilla, and 1 cup of graham cracker crumbs. However, she only has two eggs left. If she does not buy eggs, how should she adjust the quantities? [May 1998 (1)]

MAIN COURSES—*Servings of robust proportions*

24.4. *Part Way.* What common fraction is one-third of the way from 1/4 to 1/3? [May 1996 (10)]

24.5. *Beanstalk Climb.* Jack climbed up the beanstalk at a uniform rate. At 2:00 p.m., he was one-sixth of the way up, and at 4:00 p.m., he was three-fourths of the way up. What fractional part of the entire beanstalk had he climbed by 3:00 p.m.? [Mar. 1996 (9)]

DESSERTS—*Delectable morsels of the right proportion*

24.6. *Three in a Ratio.* If $a{:}b{:}c = 3{:}1{:}5$, what is the value of

$$\frac{(2a + 3b)}{(4b + 3c)}?$$

[Jan. 1996 (20)]

24.7. *Brake Job.* When you have 53 000 miles on your car, your mechanic tells you that your front brakes—your first set—are 50 percent gone and that they should be replaced when 20 percent remains. Assuming constant wear, what will your mileage be when they need to be replaced? [May 1998 (7)]

SOLUTIONS TO PROBLEMS ABOUT PROPORTIONS

24.1. *Jose's Shadow.* 24 feet. The tree's shadow is four times as long as Jose's shadow, so the tree should be four times as tall as Jose. Four times 6 feet is 24 feet.

24.2. *Zips and Zaps.* 300 pounds. Set up a table.

Zip	Zap	Zowwie	Swoosh	Pounds
1	3			C
	2	5		B
		3	2	A
			1	60

Use the bottom two rows to find $A = 120$, the next two rows up to find $B = 200$, and then the top two rows to find $C = 300$.

24.3. *Cathy's Cheesecake.* Use 2/3 of all quantities: 2 packages of cream cheese, 1/2 cup sugar, 2 eggs, 2/3 teaspoon vanilla, and 2/3 cup graham cracker crumbs.

24.4. *Part Way.* 5/18. The distance from 1/4 to 1/3 is $1/3 - 1/4 = 1/12$. Next, find one-third of 1/12 to get 1/36. Finally, add the 1/36 to 1/4 to get 10/36, which can be reduced to 5/18.

24.5. *Beanstalk Climb.* 11/24. Since Jack climbed at a uniform rate, the distance traveled from 2:00 to 3:00 should be the same as the distance traveled from 3:00 to 4:00. Jack's location at 3:00 should be halfway between his positions at 2:00 and 4:00. The midpoint of 1/6 and 3/4 is $(1/2)(1/6 + 3/4) = (1/2)(11/12) = 11/24$.

24.6. *Three in a Ratio.* 9/19. The smallest numbers that satisfy the conditions are $a = 3$, $b = 1$, $c = 5$. Substituting into the given expression results in $2a + 3b = 9$ and $4b + 3c = 19$. You can also try $a = 6$, $b = 2$, $c = 10$ and get 18 /38. A second solution is to note that $a = 3b$ and $c = 5b$. Then, $2a + 3b = 6b + 3b = 9b$ and $4b + 3c = 4b + 15b = 19b$. So the fraction is always $9b/19b = 9/19$.

24.7. *Brake Job.* 84 800 miles. Your brakes were 50 percent gone at 53 000 miles. Therefore, they were 10 percent gone at 53 000/5 = 10 600 miles. They should be 80 percent gone at $10\,600 \times 8 = 84\,800$ miles.

25

Menu for Problems about Races

APPETIZER—*A light race to whet your competitive spirit*

25.1. *Elevator Race.* Two elevators leave the sixth floor at 2:00 p.m. The faster elevator takes one minute between floors, and the slower elevator takes two minutes between floors. The first elevator to reach a floor must stop for three minutes to take on passengers. Which elevator reaches the lobby on the first floor first? [Apr. 1994 (4)]

MAIN COURSES—*Robust fare for thoughtful effort*

25.2. *Frog Race.* Two frogs have a race. One frog makes a jump of 80 centimeters once every five seconds. The other frog makes a jump of 15 centimeters every second. The rules of the race require that the frogs must cross a line 5 meters from the start line and then return to the start line to complete the race. Which frog wins the race? [Dec. 1994 (12)]

25.3. *Fox and Hound.* A fox is forty leaps [that is, fox leaps] ahead of a hound and makes three leaps while the hound makes two. Two of the hound's leaps cover as much ground as four of the fox's. How many leaps will the hound take to catch the fox? [Oct. 1997 (14)]

25.4. *Amazing Sprinters.* Three amazing sprinters run around a circular track that is 400 yards around. Sonic the Hedgehog and the Road Runner begin at the starting line and run in opposite directions, Sonic going clockwise and the Road Runner, counterclockwise. The Tasmanian Devil begins at a point exactly half way around the track and runs clockwise. They all start at the same time. Each time two runners meet, they immediately turn around and run the other way. Each of them runs at a constant speed of fifty yards a second, even while changing directions. How long will it take before the thirtieth meeting occurs? Which runners will meet at the thirtieth meeting? [Jan. 1998 (15)]

DESSERTS—*Delectable races for the spirited problem solver*

25.5. *Three-Mile Race.* A runner in a three-mile race averaged 6 mph for the first mile, 5 mph for the second mile, and 4 mph for the third mile. How long did it take the runner to complete the race? [Mar. 1997 (6)]

25.6. *Joggers Meet.* Two joggers are running around an oval track in opposite directions. One jogger runs around the track in 56 seconds. The joggers meet every 24 seconds. How many seconds does it take for the second jogger to run around the track? [Apr. 1997 (7)]

SOLUTIONS TO PROBLEMS ABOUT RACES

25.1. *Elevator Race.* The slower one. (Make a time chart of the location of the elevators for each minute.)

Time	2:00	:01	:02	:03	:04	:05	:06	:07	:08	:09	:10	:11	:12	:13	:14
Fast	6	5	5	5	5	4	3	3	3	3	2	2	2	2	1
Slow	6	–	5	–	4	4	4	4	–	3	–	2	–	1	–

25.2. *Frog Race.* The frog with the shorter jumps wins the race by two seconds. The first frog must take seven jumps of 80 cm—totaling 5.6 meters—before turning around, then take seven jumps back. The fourteen jumps would take 14 × 5 seconds = 70 seconds. The second frog must take thirty-four jumps of 15 cm—totaling 5.1 meters—before turning around, then take thirty-four jumps back. The sixty-eight jumps would take 68 seconds.

25.3. *Fox and Hounds.* Eighty hound leaps (or 160 fox leaps). Let F represent the length of a fox leap and H represent the length of a hound leap. We know that $2F = H$. Think of the time it takes the hound to cover two hound leaps as one "tick" of a timer. Then, in each tick, the hound gains a length equal to one fox leap (or one-half hound leap). The hound needs to gain 40 fox leaps; therefore, it will take 40 ticks to catch the fox. In 40 ticks, the hound will cover 80 hound leaps, which is 160 fox leaps. In 40 ticks, the fox will cover 120 fox leaps. The net gain for the hound is 160 – 120 = 40 fox leaps.

25.4. *Amazing Sprinters.* In 60 seconds, Road Runner and Sonic will meet. Let A denote the start point, B denote the 100-yard mark going counterclockwise, C denote the 200-yard mark opposite the start point, and D represent the point opposite B. Record the first several meetings.

first meeting:	RR and T at point B after 2 seconds
second meeting:	T and S at point C after 4 seconds
third meeting:	S and RR at point D after 6 seconds
fourth meeting:	RR and T at point A after 8 seconds
fifth meeting:	T and S at point B after 10 seconds
sixth meeting:	S and RR at point C after 12 seconds.

The thirtieth meeting will occur after 60 seconds. There is a cycle that repeats every eighth meeting; so the thirtieth meeting will be the same as the sixth meeting.

25.5. *Three-Mile Race.* Thirty-seven minutes. One mile at 6 mph takes 1/6 hour or 10 minutes; at 5 mph, it takes 1/5 hour or 12 minutes; and at 4 mph, it takes 1/4 hour or 15 minutes. The total time is 10 + 12 + 15 = 37 minutes or 1/6 + 1/5 + 1/4 hours.

25.6. *Joggers Meet.* Forty-two seconds. In 24 seconds, the first jogger is 24/56, or 3/7, of the way around the track. Since they have met at that point, the second jogger has covered 4/7 of the track. Since the second jogger takes 24 seconds to cover 4/7 of the track, 1/7 of the track could be covered in 6 seconds, and the complete track could be covered in 42 seconds.

26

Menu for Problems about Square Figures

MAIN COURSES—*Robust problems involving square figures*

26.1. Two Squares. Draw two additional squares in the diagram at right so that all of the stars are separated. [Nov. 1996 (15)]

26.2. Good Numbers. If a square can be partitioned into *n* squares, then *n* is called a "good number." As the diagrams show, 1, 4, 9, and 7 are good numbers. Find all positive whole numbers that are not good numbers. [Apr. 1997 (11)]

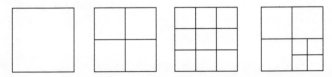

26.3. Domino. A rectangle consists of two squares placed side by side. The perimeter of the rectangle is 60 centimeters. What is the area of the rectangle? [May 1997 (14)]

26.4. Half Square. A square is folded in half to form a rectangle. If the resulting rectangle has a perimeter of twelve inches, what is the area of the original square? [Feb. 1998 (11)]

DESSERTS—*Delectable morsels (for cubes)*

26.5. Jacket for Cube. Six squares can be connected in thirty-five ways such that they all share at least one common side. Which of those combinations can be folded to make a cube ? (See example in margin.) [Jan. 1995 (13)]

26.6. Overlap Squares. Draw a diagram with three squares that overlap in such a way that six squares are formed. Draw a diagram with four squares that overlap in such a way that seven squares are formed. [Mar. 1995 (18)]

26.7. Squares and Rectangles. Nine squares are arranged in a 3 × 3 array. Embedded in the figure are a number of rectangles, some of which are also squares. What fractional part of the total number of rectangles are squares? [Mar. 1997 (7)]

26.8. Geoboard Squares. Twenty-five dots are arranged in a 5 × 5 square array. How many squares can you draw using dots as vertices? Squares may overlap. [Dec. 1997 (8)]

Figure for problem 26.1

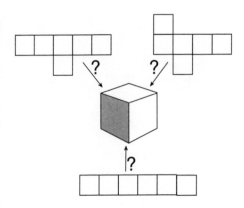

Figure for problem 26.5

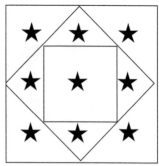

Figure for solution 26.1

SOLUTIONS TO PROBLEMS ABOUT SQUARE FIGURES

26.1. Two Squares. See figure in margin.

26.2. Good Numbers. 2, 3, and 5. If any number is good, then 3 plus that number is good because any existing square can be partitioned into four squares for a net gain of $4 - 1 = 3$ squares. So if 6, 7, and 8 are good, then all numbers greater than 8 are good. Examples for 6, 7, and 8 are shown. Since 1 and 4 are good, that leaves only 2, 3, and 5 that are not.

26.3. Domino. 200 square centimeters. If the perimeter of the rectangle is 60 cm, then the rectangle must be 20 cm by 10 cm. So the area is $20 \times 10 = 200$.

26.4. Half Square. Sixteen square inches. Represent the length of a side of the original square by x. The perimeter of the rectangle formed when the square is folded is $x + x + x/2 + x/2 = 3x$. Since the perimeter of the rectangle is 12, we know $3x = 12$, so $x = 4$. The original square is 4×4 and has an area of 16 square inches.

26.5. Jacket for Cube. There are eleven such arrangements (see margin).

26.6. Overlap Squares.

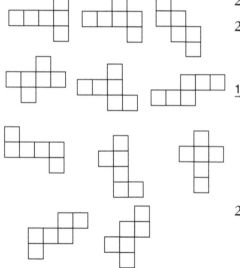

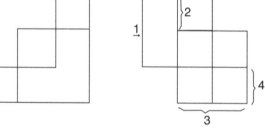

Figure for solution 26.5

26.7. Squares and Rectangles. 7/18. The figure includes the following rectangles: nine that are 1×1, four that are 2×2, one that is 3×3, twelve that are 1×2, six that are 1×3, and four that are 2×3. Fourteen ($1 + 4 + 9$) of the thirty-six rectangles and squares are squares. The fractional part is $14/36 = 7/18$.

26.8. Geoboard Squares. Fifty squares. There are thirty squares with horizontal and vertical sides as follows: sixteen of size 1×1, nine of size 2×2, four of size 3×3, and one of size 4×4. There are 20 squares with sides built on the hypotenuse of a right triangle: nine built on the hypotenuse of a 1×1 right triangle, four built on the hypotenuse of a 1×2 right triangle, one built on the hypotenuse of a 1×3 right triangle, four built on the hypotenuse of a 2×1 right triangle, one built on the hypotenuse of a 2×2 right triangle, and one built on the hypotenuse of a 3×1 right triangle.

27

Menu for Problems about Square Numbers

APPETIZERS—*Light fare of square numbers*

27.1. *Square Multiple.* What is the smallest positive integer by which 90 could be multiplied to make the product a perfect square? What is the smallest positive integer by which 90 could be multiplied to make the product a perfect cube? [Nov. 1994 (5)]

27.2. *Princess Problem.* The evil warlord, Grossout, locked Princess Stunning in the dungeon. Before he lets her out, she must find the first perfect square number larger than 100 where the sum of the digits is not a perfect square. Help the princess find the number. [Apr. 1996 (1)]

MAIN COURSES—*Mental food for thinking about square numbers*

27.3 *Four-Digit Square.* A certain four-digit perfect square is formed when two two-digit numbers are placed next to each other. If the number represented by the first two digits is one more than the number represented by the last two digits, and if the number represented by the last two digits is a perfect square, find the four-digit number. [Jan. 1997 (14)]

27.4 *Count Squares.* Six numbers between 3 and 30 are either squares or cubes: 4, 8, 9, 16, 25, and 27. How many numbers between 30 and 300 are either squares or cubes? Note: If a number is a square and a cube, count it only once. [Mar. 1997 (10)]

DESSERTS—*Delectable morsels of square numbers*

27.5 *Hypatia Numbers.* Hypatia was a female mathematician born about 370 A.D. One problem she posed was the following: Find a number that is the sum of two squares and whose square is also the sum of two squares. She looked for solutions in the class of numbers of the form $4n + 1$, where $n = 1$, 2, 3, … What numbers from this pattern yield a solution? [Apr. 1994 (19)]

27.6. *Square Dance.* Sally invited seventeen guests to her party. When it came time for the "square dance," she assigned each guest a number from 2 through 18, keeping 1 for herself. Everyone had to have a partner so that the sum of each couple's number was a perfect square. What was the number of Sally's partner? [Mar. 1996 (18)]

27.7. *Three Squares.* The number of days in a year, 365, is a special number because it is the sum of three squares. Find three squares that add to 365 to show that 365 is special. [Nov. 1996 (6)]

27.8. *Three Squares Again.* The number of days in a leap year is 366. Is 366 a special number that can represented as the sum of three squares? [Nov. 1996 (7)]

SOLUTIONS TO PROBLEMS ABOUT SQUARE NUMBERS

27.1. *Square Multiple.* 10. Since $90 = (2)(3)^2(5)$; we need factors of 2 and 5 to create the next square multiple of 90. To make a cube, we would need two factors of 2, one factor of 3, and two factors of 5; we need $2^2 \times 3 \times 5^2 = 300$.

27.2. *Princess Problem.* 256. List perfect squares greater than 100 and check the sums of digits until you get a digital sum that is not a square.

number	121	144	169	196	225	256
digital sum	4	9	16	16	9	13

27.3. *Four-Digit Square.* 8281. Consider all of the two-digit perfect squares: 16, 25, 36, 49, 64, and 81. Because the first two digits are one more than a two-digit square, the possible answers are 1716, 2625, 3736, 5049, 6564, and 8281. The only square in the list is 8281.

27.4. *Count Squares.* 14. The numbers are 36, 49, 64, 81, 100, 121, 125, 144, 169, 196, 216, 225, 256, and 289.

27.5. *Hypatia Squares.* When $4n + 1$ is a prime number, then it is a solution. For example, if $n = 1$, then $4n + 1 = 5$, a prime. This is a solution since $5 = 1^2 + 2^2$ and $25 = 5^2 = 3^2 + 4^2$. Check to see that when $n = 3$, then $4n + 1 = 13$ is a prime. Check to see that 13 and its square are both the sum of squares. *Note:* There are solutions that are not of the form $4n + 1$; 10 is a solution but is not of the form $4n + 1$. In a letter to Readers Write, Eric Siegel conjectures that "$4n + 1$ yields a solution if the number is prime or if the number is a multiple of 5." For $n = 6$, 11, and 16, $4n + 1 = 25$, 45, and 65, respectively, and each of these is also a solution.

27.6. *Square Dance.* 15. If the sum of two distinct numbers from 1 through 18 is a square, then that square must be 4, 9, 16, or 25. No pair involving 18, 17, or 16 can sum to 4, 9, or 16. Therefore, (18, 7), (17, 8), and (16, 9) must be paired. The possible pairings that remain are—for 25: (15, 10), (14, 11), and (13, 12); for 16: (15, 1), (14, 2), (13, 3), (12, 4), (11, 5), and (10, 6); for 9: (6, 3) and (5, 4); and for 4: (3, 1). Since 2 appears only once, (14, 2) must be a pair, which eliminates (14, 11) and makes (11, 5) the only pair with 11. Continuing in this manner, we eliminate (5, 4), which forces (12, 4) to be a pair. This eliminates (13, 12) and forces (13, 3). This eliminates (3, 1) and forces (15, 1). The remaining pair is (10, 6).

27.7. *Three Squares.* There are four ways to write 365 as the sum of three squares: $324 + 25 + 16$, $256 + 100 + 9$, $196 + 144 + 25$, and $144 + 121 + 100$. One strategy is to list the squares from 1 to 361. Find the first trio in the list by starting with 324 as one potential addend. Then represent $365 - 324 = 41$ as the sum of two squares; 25 and 16 are the only possibilities. Then move to 289 as the next largest possible addend and continue with this strategy.

27.8. *Three Squares Again.* Yes, 366 can be represented as the sum of three squares in three different ways. $366 = 361 + 4 + 1 = 196 + 169 + 1 = 196 + 121 + 49$.

28
Menu for Problems about Sums I

APPETIZERS—*Light fare to add to your appetite*

28.1. *Goldbach Sum.* Christian Goldbach once conjectured that any even number greater than 2 could be written as the sum of two prime numbers. Can you show that Goldbach's conjecture is true for the even numbers less than 101? [Sept. 1994 (5)]

28.2. *Sum to 100.* Do any numbers from the following set have a sum of 100? If so, which numbers sum to 100? If not, explain why not. {3, 6, 12, 21, 27, 42, 51} [May 1995 (4)]

MAIN COURSES—*Robust problems to satisfy your added hunger*

28.3. *Sum of Consecutives.* Express the number 90 as the sum of two or more consecutive integers in five different ways. [Jan. 1995 (11)]

28.4. *Magic Array.* Arrange the numbers 1, 2, 3, 4, 5, 6, 7, 8, 9, 10, 11, and 12 in a 4 × 3 array so that the sum of the numbers in each column is the same. What is the sum of the numbers in each column? [Oct. 1996 (17)]

28.5. *Partitions.* The Indian mathematician, Ramanujan, was called "the man who loved numbers." He found that the number 4 could be partitioned into the following addends:

$$4 = 1 + 1 + 1 + 1$$
$$4 = 2 + 1 + 1$$
$$4 = 2 + 2$$
$$4 = 3 + 1$$
$$4 = 4$$

Find all of the partitions for the number 5. [Dec. 1997 (9)]

DESSERTS—*Delectable sums to enrich the fussy adder*

28.6. *Sum of Multiples.* Find the sum of all multiples of N from N through $100N$, where N is any whole number. [Sept. 1994 (21)]

28.7. *Infinite Sum.* What is the value of the following expression where all the whole numbers 1–99 are included and the repeated operations are add, add, subtract? [May 1997 (8)]

$$0 + 1 + 2 - 3 + 4 + 5 - 6 + 7 + 8 - 9 + \ldots + 97 + 98 - 99.$$

SOLUTIONS TO PROBLEMS ABOUT SUMS I

28.1. *Goldbach Sum.* Some examples: $4 = 2 + 2$, $6 = 3 + 3$, $8 = 5 + 3$, ..., $14 = 11 + 3$, $16 = 11 + 5$, ..., $24 = 17 + 7$, $26 = 19 + 7$, ..., $44 = 41 + 3$, $46 = 43 + 3$, ..., $96 = 53 + 43$, $98 = 19 + 79$, $100 = 3 + 97$. There are examples for all even numbers.

28.2. *Sum to 100.* No. All members of the set are multiples of three. Any sum will be a multiple of three, but 100 is not a multiple of three.

28.3. *Sum of Consecutives.* $29 + 30 + 31$, $21 + 22 + 23 + 24$, $16 + 17 + 18 + 19 + 20$, $6 + 7 + 8 + ... + 13 + 14$, $2 + 3 + ... + 12 + 13$.

28.4. *Magic Array.* 26. The sum of the first twelve positive integers is 78. If they are divided into three equal sums, each sum must be $78/3 = 26$. One way to divide the numbers is to put 1, 2, 11, and 12 in one column; 3, 4, 9, and 10 in another column, and 5, 6, 7, and 8 in a third column. Can you find other ways to arrange the numbers?

28.5. *Partitions.* Seven partitions—5: $4 + 1$, $3 + 2$, $3 + 1 + 1$, $2 + 2 + 1$, $2 + 1 + 1 + 1$, and $1 + 1 + 1 + 1 + 1$.

28.6. *Sum of Multiples.* 5050N. Note that $N + 2N + 3N + 4N + ... + 100N = N(1 + 2 + 3 + ... + 100) = N(1 + 100 + 2 + 99 + 3 + 98 + ... + 49 + 51 + 50) = N(49 \times 101 + 50) = N(5050)$.

28.7. *Infinite Sum.* 1584. Look at the first few terms to suggest a pattern. It appears that every multiple of 3 is subtracted. Grouping three terms at a time results in $(1 + 2 - 3) + (4 + 5 - 6) + (7 + 8 - 9) + ... + (97 + 98 - 99) = 0 + 3 + 6 + 9 + ... + 96 = 3(1 + 2 + 3 + ... + 32)$. Since the sum of the first N consecutive positive whole numbers is $N(N + 1)/2$, the sum of the first 32 whole numbers is $32(33)/2 = 528$. So, the sequence has value $3 \times 528 = 1584$. Alternatively, you can group the sequence as $1 + (2 - 3) + 4 + (5 - 6) + 7 + (8 - 9) + ... + 97 + (98 - 99) = (1 + 4 + 7 + ... + 97) + (-1)(33) = 98(33)/2 - 33 = 1584$.

29

Menu for Problems about Sums II

APPETIZERS—*Light fare of more sums*

29.1. Sum of Primes. Express the number 24 as the sum of two primes in as many ways as possible. [Mar. 1996 (2)]

29.2. Consecutive Odds. The sum of five consecutive odd numbers is 65. Find the least of these numbers. [Sept. 1996 (1)]

29.3. Missing Digit. Letters have replaced some digits in the expression $a4b + 2c5 = d1e$. Given that each digit 1 through 9 appears exactly once, which digit should replace the letter b? [May 1997 (3)]

MAIN COURSES—*Mental food to sustain the efforts of adders*

29.4. Odd Sum. The sum of the first 100 positive whole numbers is 5050. What is the sum of the first 100 positive odd whole numbers? [Dec. 1996 (12)]

29.5. Alternating Sequence. Consider the infinite sum $96 + 48 + 24 + 12 + \ldots$, where each addend is half of the previous addend and the pattern continues without end. The sum of the first ten terms is 191.8125. Each addend gets you midway from where you are to 192, so the sum gets as close as you want to 192. Next, consider the sum $72 + (-36) + 18 + (-9) + 4.5 + \ldots$, whose addends are alternately positive and negative and where each addend is $(-1/2)$ times the previous addend. This infinite sum is some whole number. Find that whole number. [Mar. 1997 (9)]

29.6. Digital Sum. What is the sum of the digits of the natural numbers from 1 to 1 million? [Oct. 1998 (9)]

DESSERTS—*Delectable enrichment for adders*

29.7. Get to 100. Use numbers from the following list no more than once to obtain a sum of 100. Find as many ways to get 100 as you can. {72, 63, 53, 51, 42, 36, 33, 17, 15, 9, 6, 3} [May 1997 (10)]

SOLUTIONS TO PROBLEMS ABOUT SUMS II

29.1. *Sum of Primes.* 11 + 13 = 7 + 17 = 5 + 19 = 24. Consider the possible primes: 2, 3, 5, 7, 11, 13, 17, 19, and 23.

29.2. *Consecutive Odds.* 9. The middle number should be 65/5 = 13. The numbers would then be 9, 11, 13, 15, and 17.

29.3. *Missing Digit.* 3. Since no repetition occurs, b must be 3, 6, 7, 8, or 9. Similarly, e must be either 3, 6, 7, 8, or 9. But from addition $b + 5 = e$ or $b + 5 = 10 + e$. Therefore, either $b = 3$ and $e = 8$ or $b = 8$ and $e = 3$. The first choice works; the second does not.

29.4. *Odd Sum.* 10 000. If you list the numbers 1, 3, 5, 7, 9, … 191, 193, 195, 197, 199, you can see they can be paired (1, 199); (3, 197); (5, 195); … (99, 101). Each pair has a sum of 200 and there are 50 such pairs.

29.5. *Alternating Sequence.* 48. You can try to evaluate the sum when two, three, four, five, six, et cetera terms are involved and make a guess. There is a formula for the sum of an infinite geometric sequence whose common ratio is, in absolute value, less than 1. It is $S = a/(1 - r)$, where a is the first term and r is the common ratio. For the given sequence, $a = 72$ and $r = -1/2$. So, $S = 72(1 - (-0.5)) = 72/(1.5) = 48$.

29.6. *Digital Sum.* 27 000 001. Consider the digits in the units place. They repeat the pattern 1, 2, 3, 4, 5, 6, 7, 8, 9, 0, and this pattern repeats itself 100 000 times going from 1 to 1 million. Therefore, the sum of the unit's digit would be 100 000 times 45 (the sum of the digits) or 4 500 000. A similar pattern exists for the ten's digits, except they occur in groups of 10 zeros, 10 ones, 10 twos, and so on, up to 10 nines. The pattern repeats 10 000 times, but the sum of one pattern is 450 = 45 × 10. Therefore, the sum of the ten's digit is the same as the sum of the unit's digits, 4 500 000. The digital sum is the same for any digit up to the 100 000's digits, so the total is 4 500 000 times 6 (for each of the six places) or 27 000 000. Then the number 1 000 000 adds another 1 to the sum.

29.7. *Get to 100.* 53 + 17 + 15 + 9 + 6 = 100. Start with 72 and look for the other addends. Since 63, 53, 51, 42, 36, and 33 all take us past 100, try 17 next. Since 72 + 17 = 89, we need 11 more, and no combination can get us 11. So try 15 instead. Since 72 + 15 = 87, we need 13 more, and none of the remaining numbers can get us to 13. Since the only untried numbers are 9, 6, and 3, and they sum to 18, we cannot get to 100 by starting at 72. Use the same reasoning, starting with 63 instead of 72. After exhausting all possibilities that start with 63, try to create the sum where 53 is the largest addend. Continue in this manner.

30

Menu for Problems about Triangles

APPETIZER—*A light snack to whet your triangle-finding skills*

30.1. *Triangles in Logo.* How many triangles can you find in the NCTM logo? [Mar. 1998 (1)]

MAIN COURSES—*Robust problems to sustain triangle-finding efforts*

30.2. *Embedded Triangles.* How many triangles are contained in the figure at right? [Sept. 1994 (11)]

30.3. *Prime Perimeter.* The length of each side of a scalene triangle ABC is a prime number. Its perimeter is also a prime number. Find the smallest possible perimeter. [Apr. 1996 (7)]

30.4. *Largest Perimeter.* The lengths of two sides of a triangle are 5 and 11. If the length of the third side is also a whole number, what is the largest possible perimeter that the triangle can have? [Mar. 1997 (11)]

30.5. *Making Triangles.* How many triangles can be made using as vertices six points arranged in a 3 × 2 array? [Sept. 1998 (6)]

30.6. *Toothpick Triangles.* Arrange three toothpicks to form a triangle. Continue the process with four, five, six, and up to twelve toothpicks. For each number of toothpicks, decide whether any triangles are possible. If so, sketch the triangle and then see if any other possible triangles can be made with the same number of toothpicks. [Nov. 1997 (12)]

DESSERTS—*Delectable morsels to enrich the triangular spirit*

30.7. *Packing Triangles.* Each side of an equilateral triangle measures ten feet. What is the number of smaller equilateral triangles, each of whose sides measures 1 foot, that would be needed to completely cover the inside of the larger triangle? [Feb. 1996 (20)]

30.8. *Pentagram.* A pentagram (5-pointed star) is inscribed in a regular pentagon. How many triangles are formed? [Oct. 1998 (11)]

Figure for problem 30.1

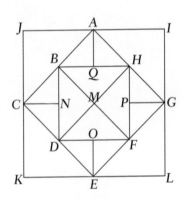

Figure for problem 30.2

Figure for problem 30.5

SOLUTIONS TO PROBLEMS ABOUT TRIANGLES

30.1. Triangles in Logo. Fourteen triangles. There are one large, four medium, and nine small triangles.

30.2. Embedded Triangles. Twenty-four triangles. Move from the smallest inner triangle outward in an organized manner. Four triangles are like *HFM*; four are like *HFD*; eight are like *GPF*; four are like *GFH*; four are like *LEG*.

30.3. Prime Perimeter. 23. List the first few prime numbers: 2, 3, 5, 7, 11, 13, 17, 23, and 29. Notice that 2 cannot be a side because the other two sides are odd primes and the sum of the three would be even and greater than 2, hence not a prime. Use guess-check-revise, starting with 3, 5, and 7, whose sum is 15 and not prime. Try 3, 5, and 11. The sum is 19, which is prime, but $3 + 5 < 11$, so no triangle can be formed. Try 5, 7 and 11. The sum is 23, and the sum of any two sides is greater than the third side.

30.4. Largest Perimeter. 31. Since each side must be less than the sum of the other two sides, the largest number that the third side could be is 15. The perimeter is $5 + 11 + 15 = 31$.

30.5. Making Triangles. Eighteen different triangles. There are twenty ways to choose three of the points from the array of six. However, two of these ways result in a collinear set, from which a triangle can not be made.

30.6. Toothpick Triangles. The following eighteen triangles can be made: 1–1–1, 2–2–1, 2–2–2, 1–3–3, 2–2–3, 2–3–3, 1–4–4, 2–3–4, 3–3–3, 2–4–4, 3–3–4, 1–5–5, 2–4–5, 3–3–5, 3–4–4, 2–5–5, 3–4–5, 4–4–4.

30.7. Packing Triangles. 100. The number of triangles in the ten rows follow the pattern: 1, 3, 5, 7, 9, 11, 13, 15, 17, 19. The sum is $1 + 19 + 3 + 17 + 5 + 15 + 7 + 13 + 9 + 11 = 5 \times 20 = 100$.

30.8. Pentagram. Thirty-five triangles. There are ten individual triangles arranged around the center pentagon. Then, there are ten more triangles made by combining two adjacent smaller triangles. Then, there are five triangles made by combining three adjacent smaller triangles. There are five triangles made by using the middle pentagon and four smaller triangles. There are five triangles made by using the center pentagon and two smaller triangles. So, there are $10 + 10 + 5 + 5 + 5 = 35$.

A Brief Introduction to Problem Posing

Problem posing is something that should naturally occur at each of the stages of solving a problem. In the ENTRY stage, you are focused on trying to understand the problem posed to you. That will invariably lead you to reformulate the given problem using terms that you understand. During the ATTACK, stage you will be formulating conjectures related to the problem to be solved. Each of these conjectures is a statement to prove or disprove and represents another problem. During this stage, you may also try to formulate and solve a simpler problem; one whose solution you hope will lead to a solution of the more difficult problem. During the REVIEW stage, you will try to formulate and to solve related problems or more general problems.

The act of posing problems is a natural part of problem solving. Becoming more adept at posing problems can enhance one's problem solving abilities. Some portion of the time devoted to problem solving and the mathematics curriculum should be spent on problem posing. To encourage those efforts, the appendixes to this publication contain some suggestions for posing problems related to the menu problems.

Some of the menu problems are situations in which you are asked to find a number or numbers that meet certain conditions. Sometimes the conditions concern the digits; sometimes the factors; sometimes a relationship to another number. To pose a problem of that type, you only need to start with a number and then try to describe that number in several ways. Next, see if the number you have in mind can be recovered from the descriptive statements. Sometimes you will find that more than one number meets the conditions, and you can choose to add another condition or choose to have a problem with multiple solutions. Experience in making up these kind of problems improves mathematical vocabulary and provides insight into the strategy of working backward. It is also possible to develop more tolerance for what you may think are poorly posed problems. If you pose a problem and others blame you for their failure to understand that problem, you will come to understand the need to be as precise as you can with the language and also come to realize that no one formulation of a problem is likely to be interpreted the same by all students who attempt to understand and to solve it.

Some appetizer menu problems that are "find a number" situations are the following: 1.1 Hot Dog Weight, 7.1 People on the Bus, 8.2 Missing Weight, and 14.1 Mystery Number. Let's try to pose another hot dog-weight problem. We begin by choosing a number for the answer—let's make it an 81-pound hot dog. Then we make three guesses. Suppose we choose 65, 108, and 94. One is off by $81 - 65 = 16$, another is off by $108 - 81 = 27$, and the third is off by $94 - 81 = 13$. Now we can pose a similar problem as follows:

Abigail, Benny and Chip tried to guess the weight of a giant hot dog. Abigail guessed 65 pounds, Benny guessed 108 pounds, and Chip guessed 94 pounds. One estimate was off by 13, another by 16, and the third by 27. What was the weight of the hot dog?

When we try to solve this problem by the strategy described in the solution to 1.1, we construct a table in which the only common entry in each of the three rows is 81, the number we started with. We have a new problem for someone else to solve, and it has a unique solution.

Many of the menu problems are situations that can be varied by asking and answering "What if?" or "What if not?" To pose a new problem, just change one or more of the given conditions. As you scan the appendixes at the back of this book, you will see a number of suggestions for creating new problems by answering the "What if?" question. Some main-course problems which contain "What If?" suggestions are 6.7 Missing Score, 10.4 Angle at 2:48, 18.5 Wait in Line, and 21.6 Odd Triangle.

Let's use the strategy to pose a problem like that in 6.7 Missing Score. In that situation, there were *five* tests, ranging from *0 to 100,* and Johnny had an average of *88.* The problem asks for the *lowest* score Johnny could get on any one test. There are four important elements that have been italicized. Any one or combination of these elements could be changed. What if we changed the number of tests? What if we change the range? What if we change Johnny's average? What if we ask for his highest score or a score other than his lowest?

Just making an arbitrary choice, suppose we make the situation as follows: there are *six* tests, ranging from *20 to 96,* and Johnny had an average of *88.* What is *lowest* score Johnny could get on any one test? We try to find a solution by reasoning that Johnny's total score was $6 \times 88 = 528$. If he got as high as possible on five tests, he would get $5 \times 96 = 480$. So the remaining test could be $528 - 480 = 48$. What if we asked for the greatest difference between his highest and lowest score?

Let's also illustrate the strategy with 21.6 Odd Triangle. The problem asks us to look for a pattern in the differences between the first and last number in a row of a triangle composed of odd numbers. It also asks for a pattern in the sum of the numbers in a row of that same triangle. The elements that can be varied are the patterns to attend to, the numbers in the array, and the shape of the array. The easiest element to vary is probably the numbers in the array. If not odd numbers, then what? Even numbers is the most obvious choice. But you could also use odd numbers that start at 3, or you could use even numbers that start at 10. You could use multiples of 3 and start where you like. But you could also look for different patterns. For example, you could try to describe the average of each row in terms of its row number.

You are encouraged to be as creative as you can be in using the "What if?" strategy to extend your experience with problem solving. The suggestions in the appendixes should help, but you should also feel free to create many more interesting problems. As a result, you will not only gain valuable experience as a problem solver but will have some insight into where problems come from.

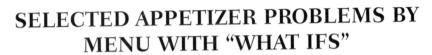

Appendix 1

SELECTED APPETIZER PROBLEMS BY MENU WITH "WHAT IFS"

1.1. *Hot Dog Weight.* This is an example of what some would call an unreasonable problem in the sense that you would never know how much each estimate was off unless you knew the weight of the hot dog in the first place. Rather than a problem, consider it a puzzle or a challenge. You can make problems of this type easily. Start with a solution—the weight of the hot dog. Next, make up estimates and figure out how far each is off. Then, make up your own problem. Can you make up a problem like this where there is more than one correct solution or where there is no solution?

1.2. *Egg Timer.* With a 7-minute timer and an 11-minute timer, could you time a 16-minute egg? A 20-minute egg? What times are possible? What times are impossible? What if you had two different timers—say, a 6-minute timer and a 10-minute timer?

1.3. *Special Number.* Can you make up other ten-digit numbers that use all digits and then describe it so that it differs from all others? Can you challenge your friends to find the description you have in mind?

1.4. *Average to 1000.* Can you find the average of whole numbers from 1 to 500 or from 1 to some other number? Can you find the average of whole numbers from 500 to 1000? What about the average of the numbers from 600 to 1200? There are lots of problems here.

1.5. *Adding Red Marbles.* If you start with a container having marbles in the ratio of 3 red for every 5 green, what ratios can you get by adding marbles? What ratios can you get by subtracting marbles?

1.6. *Painting Pair.* Fair-share problems are very real and have been for generations. Make up your own problem by changing the amount of time that two people work and the total amount paid for the complete job.

1.7. *Notebook Cost.* What if four notebooks were purchased? Could the price of one notebook be determined? What if the change were all in dimes? Could the price of one notebook be determined?

1.8. *True or False.* These are fun to make up. Look up some unit of measure and write a meaningful sentence and a ridiculous sentence using that measure. Then see who can tell which is which. Can you run a furlong in ten seconds?

7.1. *People on the Bus.* This is another one of those problems that cannot be stated sensibly unless you know the answer. To construct another problem, start by making up an answer and then working backward. Is there an advantage to starting with a number that has several divisors?

7.2. *Reds and Greens.* What if the number of marbles in the container is changed? What if the ratio of the colors in the original container is changed? What if you want to add marbles to get a different ratio than 1/2? What if you were to take away marbles from the container? Could you add both red and green marbles to the container and get a solution?

8.1. *Nine Coins.* What if you varied the number of coins? Why was the number "9" chosen for this problem?

8.2. *Missing Weight.* This problem and others like it are designed to introduce students to problems that can be solved by setting up a system of two linear equations. However, it is expected that a natural solution will be a concrete introduction to the algebraic solution. Again, if you want to make up some more of these problems, it is convenient to start with a solution—a weight for Block A and a weight for Block B. Make up two statements of equality. Then make up a question.

9.1. *Count Prime Dates.* Can you find prime dates for other years? Do the number of prime dates differ from year to year? Which month has the most prime dates? Which month has the least prime dates? How many dates are there so that the number of the month and the number of the day are relatively prime? (That is, have no common factors other than 1.)

9.2. *Last Prime Date.* What are the first and last dates that are relatively prime dates? Which months have the fewest relatively prime dates? Which months have the most relatively prime dates?

10.1. *Fives.* What if the digital clock were to contain at least two fours? Two nines? Would there be more times when the clock would have at least two fours than at least two fives? Would there be more times when the clock would have at least two ones than at least two nines? What about the number of times at least three digits are fives?

11.1. *Different Sums.* What if you were to change the labels on the numbered cubes? Can you change the number of possible sums? What if the labels of one have to be the opposites of the labels of the other?

11.2. *Make Change.* What if you could use only dimes or pennies? What if you could use only nickels or pennies? What if you could use dimes, nickels, or pennies? What if you could use quarters as well?

12.1. *Surface Area.* What if you did not need to make rectangular prisms? How many different stackings of unit cubes can you make so that the surface area is 24?

12.2. *Pack a Box.* What if the cubes being packed were of different sizes? What if the box they were to be packed in was a different size?

13.1. *"3" in Tag.* What if we want to know the number of "sevens" used? Is there one numeral used more frequently than all others? What if the total number of tags were different from 100?

13.2. Digits on Tickets. What if there were a different number of seats in the place where the play was being staged? What if the tickets were numbered from 100 to 450?

13.3. Sevens. What if we wanted to know the fraction of numbers containing a five rather than containing a seven? What if we wanted to know about other digits?

14.1. Mystery Number. Make up a new problem by working backward. Pick a number with at least six factors. Choose three of the factors to be revealed. You may give other information like the total number of factors. See if your number can be recovered. See if more than one number meets the conditions.

14.2. Multiply Ten. What if we wanted the first eight natural numbers? What if we wanted the first twelve? Could you do the problem for other sets of numbers?

14.3. Divisors of 72. What if we wanted to know about numbers other than 72? Are there numbers, N, so that if k and m divide N, then km must also divide N?

15.1. Use All Digits. What if we wanted the largest multiple of eight? The largest multiple of twenty? The largest prime?

15.2. Unusual Age. This is another work-backward situation. Think of a number that could be an age for someone—an age that is a multiple of some other number. Look at ages 1, 2, 3, 4, and so on, years into the future and find a number that is a multiple of some other number. Ask when this would occur again.

16.1. Three Conditions. Another work-backward situation. Think of a number and make up three clues. See who can figure out your number.

16.2. A Decimal Number. Also another work-backward situation. This differs from the previous one in that the number you start with can be represented as a decimal and you can describe it using some terms that are not applicable to the previous situation. For example, you can deal with "rounding."

16.3. Two Digits. What if you were looking for three-digit numbers?

17.1. Rectangles. What if you changed the picture to a 3 × 5 array or a 4 × 7 array?

17.2. Three Straws. Can you go on to four straws?

17.3. Find Triangles. What if you drew all of the diagonals of a regular hexagon? How many triangles would you make?

18.1. Miniproduct. What if the five digits you used were 2, 3, 4, 5, and 6? What if the five digits were 3, 4, 5, 6, and 7? What if only even digits could be used? What if only odd digits can be used?

18.2. Close But Less. What if the target were a number different from 30? What if it were 25 or some other number? What if the digits you use are 2, 3, 4, 5, 6, and 7? What if there are five digits you can use? What if there are seven digits you can use?

18.3. **Just Close.** Same questions as in 18.2 except you do not have to stay under.

19.1. **Around a Wheel.** Will you always be able to roll one wheel around another wheel and return to an original position? What happens as you vary the sizes of the wheels?

19.2. **Area.** You can make up others like this by drawing two overlapping rectangles. Do you need all of the information given in the picture or can you find the area without some of the information?

19.3. **Angle Size.** What if the triangle is not equilateral? What if the triangle is isosceles and angle *P* measures 70 degrees?

20.1. **Shortcut.** Would you save proportionately more time cutting across a field that was square or one that was much longer than it was wide?

20.2. **Tree Replacement.** Can you devise another rule that would be reasonable in the sense that it is unlikely that you would be able to replace a damaged tree by a single tree?

21.1. **Nine Stones.** What if there were a different number of stones?

21.2. **Triangular Array.** What if the first number is something other than 1?

21.3. **Hundreds Chart.** Can you make up other portions of a hundreds chart and then find what number goes in a given location? What if the chart went from 1 to 12 in the first row? What if the chart went from 1 to 7 in the first row?

22.1. **Five Spades.** What if you change the number of spades required? What if you have an incomplete deck of cards—no face cards, no red cards, or no sevens?

22.2. **Boys and Girls.** What if the family has a different number of children? How does the answer change as you vary the size of the family?

22.3. **Shake a Three.** What if the dice are numbered cubes that you place different numbers than 1, 2, 3, 4, 5, and 6 on? What about rolling a multiple of five?

23.1. **Scratch Off.** What if a winner had to match three spots rather than two? How does that change the odds? What if Bob wants the probability of winning to be about 0.01? What if he wants it to be about 0.001?

23.2. **Red, Green, and Blue.** Can the problem be solved by taking marbles out of the bag instead of adding marbles? What if the number of marbles of the different colors is changed?

23.3. **Shaded Triangles.** What if you made a more complicated figure—a triangle that had been partitioned into four triangles (by drawing a segment parallel to the baseline through the point that all of the smaller triangles have in common)? What if you partitioned a triangle into nine, rather than four, triangles?

24.1. **Jose's Shadow.** What if the numbers were changed? Can you find a number of situations in which the height of the tree is a positive integer?

24.2. Zips and Zaps. Does the introduction of terms like *zip* and *zap* make the problem easier or more difficult?

24.3. Cathy's Cheesecake. How does the quantity of eggs have an effect on the variety of ways you can make this recipe?

25.1. Elevator Race. What if you change the number of floors in the building? What if you change the length of time it takes to stop and discharge passengers?

27.1. Square Multiple. What if you changed the initial number to 110 or 111 or to some other number? Can you find a number between 100 and 200 that has a fairly large number that is required as a multiplier to produce a perfect square? That has a fairly small number that is required as a multiplier to produce a perfect square?

27.2. Princess Problem. What if she must find the first such number that is greater than 150? Greater than 200?

28.1. Goldbach Sum. Can you find several even numbers which can be expressed as the sum of two primes in more than one way? Can you find a number that can be expressed as the sum of primes in more than two ways?

28.2. Sum to 100. What if you changed the numbers that are in the set? What if you tried to make sums different from 100?

29.1. Sum of Primes. This is a special case of 28.1.

29.2. Consecutive Odds. What if the number of consecutive odds is changed from five to four or to seven? What if you need consecutive evens?

29.3. Missing Digit. Can you make up several problems by finding a sum of a three-digit number and a three-digit number and then erase some of the digits and then recover them?

Appendix 2

SELECTED MAIN-COURSE PROBLEMS BY MENU WITH "WHAT IFS"

5.1. *Make 1–10.* What if you had to use 1, 2, and 5 each exactly once? What if you had to use 1, 3, and 6 each exactly once? What if you had to form numbers greater than 10?

5.2. *Palindrome Bus.* What if you wanted to find the distance to the next palindrome from a different five-digit number? What if the given number were a six-digit number?

5.5. *Jenna's Handicap.* What if the handicaps—points given at the start to Jenna and Courtney—were changed?

5.6 . *Fruit Puzzle.* Can you make up another puzzle like this?

5.7. *How Many Pizzas?* What if the basic price is changed? What if the cost of toppings is changed? What if the amount of money that Chris has is changed?

5.8. *Storing Paper.* This is a problem intended to be solved by estimation rather than by applying computation to all given data. Can you find another thing that your school has delivered and make up another estimation problem?

6.1. *Arranging Operations.* What if you changed the numerals 5, 4, 6, and 3 to some other numerals? Can you put in numbers where the number of different values you can obtain is less than with the given numbers? Where the number of different values you can obtain is more than with the given numbers?

6.2. *Difference of Squares.* What if you changed the difference or the difference of squares?

6.3. *On and Off.* What if the number of stations is changed from 20 to 18 or to some other number?

6.4. *Ball, Book, and Mug.* What if there were four objects to be purchased and you knew the cost of four pairs of the objects? Could the problem be done then?

6.5. *Novel Operator.* Can you compute the value when the input is different from 3, 5, and 8? Can you find some triples that are associative? You can make up your own novel operator and then compute the output for a given three inputs.

6.6. *Equivalent Fractions.* Are there other sets of fractions that have this same property?

6.7. *Missing Score.* What if Johnny's average were different from 88? What if the number of tests were changed?

6.8. *Idle Days.* What if the man had been hired for thirty days and received $45? What if the day's wages and/or the penalty for idleness were altered?

7.3. *Knife, Fork, and Spoon.* This is another of those "balance" situations that provide an intuitive introduction to systems of linear equations. What if you change the balances in the problem?

7.4. *Boxes inside Boxes.* What if there were five little boxes inside some bigger boxes rather than six? What if the number of medium-sized boxes is not the same as the number of little boxes? What if the total number of boxes is changed?

7.5. *Discount This.* Can you argue that the discount is not 5 percent?

8.3. *Tiling, Tiling.* What if the tiles are 3 inches square? What if the tiles are 4 inches square? What if the tiles are $1'' \times 5''$?

8.4. *Fill the Tub.* What if only the cold water and the waste pipe were operating? What would happen then? What if the tub were initially one-quarter full?

8.5. *Elf on the Stairs.* Compare this problem with the task of finding the number of ways to tile a 2×10 walkway with bricks that are 1×2.

9.3. *Product Dates.* What if you were not limited to this century? Can you find other years when there are no such dates?

9.4. *Vote Date.* What if it could be any first Tuesday and not have to be after a first Monday?

10.4. *Angle at 2:48.* What if you wanted to compute the angle at some other time?

11.3. *Toppings.* What if the restaurant claimed to have over 2000 varieties?

11.4. *Pass the Cookies.* Try to find a general way to do this distribution problem by thinking of the cookies as lined up in a row with dividing bars placed in the row to separate the cookies into three subsets. See if this works with larger numbers of cookies.

12.3. *Painted Cubes.* What numbers are possible for the number of unpainted cubes? Could you do the problem if you knew the number of cubes that had two sides painted? Could you do the problem if you knew the number of cubes that had three sides painted?

13.4. *The Last One.* What if Betty started with a number other than 1? What if the number of digits Betty wrote were changed? What if we only knew the number of times that Betty had written the digit 7?

13.5. *Magic Triangle.* Are there any other magic numbers possible?

13.6. *Add to Eight.* What if we check a different range of numbers? What if we look for a different digital sum?

14.5. *Rectangle Building*. What if Bobby has a different number of tiles? What if Bobby does not have to build rectangles but does have to place the blocks edge to edge?

14.7. *Counting Zeros*. What if we wanted terminal zeroes in 120!?

15.3. *Pricing Folders*. What if we had not known that the folders were originally priced at 50 cents? Suppose that a different amount of money had been brought in with the sale?

15.4. *Find the digits*. This is another problem that you can make up by finding a product on your calculator. Erase one digit in the product and see if you can recover the digit knowing one of the factors.

15.6. *Make a Product*. What if the product can contain more than three digits?

16.4. *Jeff's Blocks*. What if Jeff had more than 1000 blocks? This is another of the type of problems you can construct by choosing a number and then finding some remainders when dividing by selected numbers. Then see if you can reconstruct the original number from the statements about the remainders.

16.5. *Ken's Shells*. What if Ken has more than 100 shells? See suggestions about Jeff's blocks.

16.6. *Becky's Shirts*. What if you changed the proportions of white and pink shirts? What if you changed the number of green shirts?

17.4. *Big Wheel, Little Wheel*. What if you changed the circumferences of the two wheels? What if you were given the radii or diameters of the wheels instead of the circumferences? What if you compared the number of turns for a distance different than one mile?

17.5. *Change in Area*. What if the percentage increases and decreases were changed? Are there any percents for increases and decreases where the volume does not change?

17.6. *Color the Logo*. What if you tried to color figures other than an NCTM logo? What kinds of figures can be colored with two colors?

18.4. *Progressive Taxes*. What if the percent tax rate were one-half the number of thousands of dollars earned? What if the percent tax rate were two-thirds of the number of thousands of dollars earned?

18.5. *Wait in Line*. What if the coaster has fifteen cars? What if each coaster holds up to four people? What if the ride time is changed? What if the unloading and loading time is changed? What if the number of people in your group changes?

18.6. *Two Sales Plans*. What if there are other options that include some guaranteed salary plus some commission? For example, a $100 salary and an 8 percent commission? What if the estimated dollar value of sales were changed?

19.5. *Prisms in a Box*. What if the box is $5'' \times 5'' \times 5''$? What if the rectangular prisms are $3'' \times 2'' \times 1''$ or $2'' \times 2'' \times 3''$?

20.3. Partitioned Rectangle. Construct a rectangle and partition it into four rectangles. Find the areas of three of the rectangles. See if the fourth area can be recovered. Can you make a situation where there is more than one solution?

20.4. Freeway Poles. Can you find the distance between any two poles? For example, can you find the distance between the 15th and the 23rd pole? Can you find how many poles there would be in a mile?

20.5. Unit Triangle. What if the medium-sized triangle were used as a unit?

21.4. Powers. Can you find the units digit of 1999^{1999}? You can make up many problems of this type and solve them by looking for patterns.

21.6. Odd Triangle. Is there a pattern for an "even" triangle?

22.4. Bags and Balls. Can you physically carry out this activity and record the results? If you do it 100 times, how many times would you expect to get a red ball? What if you vary the number of red and green marbles in each bag?

22.5. Sum and Product. What if the numbers are selected from the set of positive integers less than or equal to 7? Or from those less than or equal to 10? What if you were to select two numbers from the positive integers up to ten and wanted their difference to be less than five? What is the probability of this event?

23.4. Toss a Tetrahedron. What if you toss the tetrahedron three times? What is the probability that the same color will be face down all three times? What is the probability that all three colors will be different?

23.5. Multiply a Pair of Dice. What is the probability of getting a product that is a multiple of four? A multiple of six? A multiple of four or six? A multiple of four and six?

23.6. Center or Edge. What if the target were a square? What if the target were an equilateral triangle? What if the target were an equilateral hexagon?

24.4. Part Way. What common fraction is one-fourth of the way from 1/3 to 3/5? What common fraction is two-thirds of the way from a/b to c/d?

24.5. Beanstalk Climb. What part of the entire beanstalk had he climbed by 2:40 p.m.? What if he was one-fifth of the way up by 2:00 p.m.? What if he was five-eighths of the way up at 4:00 p.m.? What time did Jack start? What time did Jack finish?

25.2. Frog Race. Can you make up another race in which one frog makes longer jumps less frequently than another frog who makes shorter jumps less frequently? What if the race is just straight in one direction from start to finish and does not require a turn around and return? Does this make a difference?

25.3. Fox and Hound. Can you make up another race in which one contestant starts with a lead but advances in jumps at a slower rate than the other contestant?

25.4. Amazing Sprinters. What if we change the rate at which they all run to 40 meters per second? To 80 meters per second?

26.3. *Domino.* What if the rectangle consisted of three squares placed side by side? What if the rectangle consisted of four squares placed side by side? What if the rectangle consisted of six squares placed in a 2 × 3 array? What if the perimeter was 100 centimeters?

26.4. *Half Square.* What if the original square had been folded twice and the folds were perpendicular to one another? What if the original square had been folded twice and the folds were parallel to one another?

27.3. *Four-Digit Square.* How many four-digit numbers of the form *abab* are squares? If you want to describe one of these square numbers, what else would you need to know? How many four-digit numbers of the form *abac* are squares? If you want to describe one of these numbers, what else would you need to know? How many four-digit palindromes are squares?

27.4. *Count Squares.* What if we were looking for squares or cubes between 300 and 3000? Between 50 and 2000?

28.3. *Sum of Consecutives.* What if we want to express the number 95 as the sum of consecutive integers in as many ways as possible? What about expressing other numbers as sums of consecutive integers?

28.4. *Magic Array.* What if you wanted the sums in the rows to be the same? Can you make the sums in the rows and the sums in the columns be the same? What if you used the first twelve even numbers or the first twelve odd numbers instead?

28.5. *Partitions.* Can you use the partitions of the number 4 to help you produce the partitions of the number 5?

29.4. *Odd Sum.* What is the sum of the first 100 multiples of three? What is the sum of 1 + 5 + 9 + 13 + ... + 397? What is the sum of 10 + 13 + 16 + 19 + ... + 307?

29.5. *Alternating Sequence.* What if you start from a whole number different from 72? What if each addend is (–1/3) times the previous addend? What if each addend is (–2/3) times the previous addend?

30.2. *Embedded Triangles.* What if you found embedded triangles in some other figure? Can you construct a figure that would be an interesting challenge?

30.3. *Prime Perimeter.* Find several other prime perimeter triangles. What if the triangle need not be scalene? What if the figure were a quadrilateral rather than a triangle?

30.4. *Largest Perimeter.* What if the given sides are 8 and 13 or some other numbers than 5 and 11? What if we wanted the smallest possible perimeter the triangle could have? What if the figure were a quadrilateral rather than a triangle?

30.5. *Making Triangles.* What if we used a 3 × 3 array of dots or a 2 × 4 array of dots?

30.6. *Toothpick Triangles.* How many toothpick triangles can be made with 13 toothpicks? With 14 toothpicks?

Appendix 3

SELECTED DESSERT PROBLEMS BY MENU WITH "WHAT IFS"

2.1. *Scroboscopi.* Make up another problem by starting with some number of scroboscopi with 2, 3, and 5 legs. Find the number of legs on three successive days. See if you can recover the original number of scroboscopi. What if you knew the number of legs on three different days but the days were not consecutive? Could you still find the number of scroboscopi?

2.2. *Gym Time.* What if the gym period were forty minutes long? What if twenty-eight students want to play?

2.3. *Mint, Chips, and Nosh.* What if you did not know that eight people had sampled all three? What possible numbers of people could sample all three flavors? Make up your own Venn diagram problem.

2.4. *Missing Data.* What if the median were 320? What if the remaining numbers did not have to be the farthest apart? What if the median were not one of the given eleven numbers?

2.5. *Share the Estate.* What if the estate has a value of $75 000? What if the widow gets $5 000 more than the son and daughter together?

2.6. *Partition a Square.* How many different combinations of four areas can a 10-centimeter square be partitioned into by a horizontal and vertical line? In which of these cases could you determine where the lines must be drawn to produce those areas?

3.1. *TV or Not TV.* If you pay $100 a month and you are charged 1.25 percent per month on your unpaid balance, how long will it take you to pay off a $5000 loan? Is the time midway between the times needed to pay off a loan that charges 1 percent per month interest and another that charges 1.5 percent per month interest?

3.3. *Ox and Sheep.* What if each ox eats as much as six sheep? What if the first drover put in six oxen and the other put in twenty sheep? What if the cost of the pasture were changed?

3.5. *Connie's Boats.* What if Connie made 10 percent on one sale and lost 10 percent on the other? What if Connie made 25 percent on one and lost 25 percent on the other? If Connie made 10 percent on one sale, what percent could she afford to lose on the other sale to break even? If Connie lost 25 percent on one sale, what percent does she need to gain on the other in order to break even?

3.6. Birthday Lunch. What if there were seven people who went to lunch and two of them had birthdays? What if the total bill was $64 ?

7.6. Count Off. Make up some other "count off" problems.

7.7. Test Scoring. What if 4 points are given for each correct answer and 2 points are deducted for each incorrect answer? What if your team scored some number other than 48?

8.8. Granny's Gift. What if your grandmother put $5000 into an account for you? What if the account earned 0.5 percent per month? What if the account earned 0.75 percent per month? Is earning 0.75 percent per month the same as earning 9 percent per year?

10.5. Digital Changes. What if we want to know the time after thirty-five digital changes? What if the start time is different than 6:38? For example, what about 12:42?

10.6. Increasing. What if the digits are nondecreasing? For example, you would now count 2:33 and 2:23.

10.7. Decreasing. What if the digits must be nonincreasing? For example, now you would count 3:32 and 3:22.

10.8. Angle at 5:20. What if you want to know the angle at different times? What about 6:10?

10.9. Losing Time. What if the clock uniformly loses three minutes every twenty-four hours?

11.6. Four-Digit Palindromes. What if we wanted to know the number of five-digit palindromes? What about six-digit palindromes?

11.7. Fruit Juggler. What if he drops six pieces of fruit? What if he drops ten pieces of fruit?

11.8. DJ's Choice. What if you change the number of rap, rock, alternative, oldies, and country singles? What if you had to play at least one single from each category in each ten-song set?

12.4. Painted Dice. For a start on a generalization, see 12.5.

12.6. Counting Colors. What if there were three colors of paint?

12.7. Remove Cubes. What if all eight corner cubes are removed? What if eight cubes that are neither corner cubes nor center cubes are removed?

12.8. One Cube to Three. What if one is six inches on a side and another is three inches on a side? What if the first two are five inches on a side?

13.7. Odd and Even. What if we are looking in the range from 0 to 199? In the range from 0 to 299? In the range from 100 to 999?

13.8. Number a Cube. Can a sum different from 18 be obtained?

14.8. Tip the Driver. Could the driver have received any amount different from $8.41? Could he have received $9.61 or $8.89? What if the number of coins is changed?

15.7. *Five Factors*. What about six factors? What about seven factors? What about four factors?

15.8. *All Digital Divisors*. What about all numbers 1 through 10? What about all whole numbers 1 through 11? What about 1 through 12?

16.7. *Up and Down*. What if the difference between the start and finish numbers is something other than eight? Make up a problem like this by starting with a number, increasing it by some percent, decreasing that by some percent, and then noting the difference between the first and last numbers. Now, given the percent increase and the percent decrease and the difference, can you recover the original number?

16.8. *Impossible Score*. What if the possible scores on a single throw are 0, 3, and 8? What if the possible scores on a single throw are 1, 3, and 5? What if the possible scores are 0, 2, and 6?

17.7. *Shady Triangles*. What if you want to shade five of the triangles? What if you want to shade two of the triangles? What if you change the original figure to a square partitioned into nine congruent squares and you want to shade four of the smaller squares?

17.8. *Patio Bricks*. What if the bricks are 2×5 and the frame is 10×12?

17.9. *Hidden Rectangles*. What if the figure is of N congruent rectangles packed end to end? What if the original figure is of twelve rectangles packed in a 2×6 array?

18.7. *Votes for B*. What if the total votes cast were changed? What if there were fifty-one votes? What if E was last with four votes?

18.8. *Maximize a Product*. What if a, b, c, and d are chosen from $\{2, 3, 4, 5\}$? Or from the set $\{1, 3, 5, 7\}$? Or from the set $\{0, 2, 4, 6\}$? What if you want the greatest possible value rather than the least?

19.6. *Compare Cylinders*. What if you took two sheets of paper that were the same size and cut $1'' \times 2''$ rectangles from the corners of each. On one, the 2-inch cuts are on the shorter edge of the paper, and on the other, the 2-inch cuts are on the longer edge. Would the boxes formed by folding up the sides hold the same amounts?

19.7. *Area of Pieces*. You can make up several problems like these.

19.8. *Area Ratio*. What if the ratio of length to width is changed? Can you find a solution if the rectangle has sides in the ratio 4 to 2? What about the ratio 5 to 2?

20.6. *Pizza Comparison*. What would be the diameter of a pizza that is twice the size of the 15-inch pizza?

20.7. *Circular Region*. What if the length of AB is 12?

20.8. *Unpainted Block*. What if the original block was $5 \times 8 \times 10$? Or $8 \times 8 \times 12$?

22.6. *Weighted Cube*. What if the cube is twice as likely to land with an odd digit showing?

23.7. *Terminators.* What if the numerator and denominator sets were interchanged? What if the numerator set was {2, 4, 8, 10, 13} and the denominator set was {4, 5, 8, 12, 15}?

24.6. *Three in a Ratio.* What if the ratio is 1:3:5? What if the ratio is 5:2:4? What if you look for the value of a different function of *a*, *b*, and *c*?

24.7. *Brake Job.* What if you have 42 000 miles on your car? What if your brakes are 60 percent gone? What if you have 42 000 miles on your car and your brakes are 60 percent gone?

25.5. *Three-Mile Race.* What if the runner averaged 4 mph for the first mile, 5 mph for the second, and 6 mph for the third? What if the runner averaged 4 mph for the first mile, 6 mph for the second, and 3 mph for the third?

25.6. *Joggers Meet.* What if the joggers meet every 26 seconds? Every 28 seconds? What if the first jogger runs a complete lap in 64 seconds?

26.5. *Jacket for Cube.* What if you looked at the arrangements of five squares that can be folded to make a jacket for a cube with an open top?

26.7. *Squares and Rectangles.* What if there were sixteen squares arranged in a 4 × 4 array?

26.8. *Geoboard Squares.* How many rectangles can you make so that one edge is twice as long as another edge?

27.6. *Square Dance.* What if the number of guests is different from seventeen? Can the problem be solved if Sally invites thirteen guests? Is there a number of guests for which there is more than one solution?

27.7. *Three Squares.* Find several other special numbers in the range from 300 to 399. Can you find four square numbers in this same range?

28.7. *Infinite Sum.* What if the repeated operations are add, subtract, subtract? What if the repeated operations are subtract, add, add?

29.7. *Get to 100.* What if the permitted list were different? What if the list is the same as that given except for the largest number and the smallest number?

30.7. *Packing Triangles.* What if each side of the triangle were twenty feet?

30.8. *Pentagram.* Can you find this problem elsewhere in the collection?

Index

ALPHABETICAL BY PROBLEM NAME